Climate Change Mitigation

POLICIES AND PROGRESS

OECD
BETTER POLICIES FOR BETTER LIVES

This work is published under the responsibility of the Secretary-General of the OECD. The opinions expressed and arguments employed herein do not necessarily reflect the official views of OECD member countries.

This document and any map included herein are without prejudice to the status of or sovereignty over any territory, to the delimitation of international frontiers and boundaries and to the name of any territory, city or area.

Please cite this publication as:
OECD (2015), *Climate Change Mitigation: Policies and Progress*, OECD Publishing, Paris.
http://dx.doi.org/10/1787/9789264238787-en

ISBN 978-92-64-23267-9 (print)
ISBN 978-92-64-23878-7 (PDF)

The statistical data for Israel are supplied by and under the responsibility of the relevant Israeli authorities. The use of such data by the OECD is without prejudice to the status of the Golan Heights, East Jerusalem and Israeli settlements in the West Bank under the terms of international law.

Photo credits: © photlook – Fotolia.com.

Corrigenda to OECD publications may be found on line at: *www.oecd.org/publishing/corrigenda*.

Preface

At COP 21 in Paris at the end of 2015, the global community has an opportunity to reach a new global deal on climate change that reaffirms its collective commitment to reducing greenhouse gas emissions. This agreement must send a clear signal that all countries are working towards a zero net carbon future. However, it is increasingly clear that keeping the increase in global average temperature below 2 °C will require ambitious national targets and goals to reduce emissions together with enhanced domestic policies and implementation on the ground.

As countries take steps to enhance the multilateral climate change regime, the OECD is supporting them to create more effective policies that take national contexts and policy priorities into account. In order to build trust, it is essential to transparently track the progress being made towards global and national climate objectives, increase the understanding of national policy contexts and learn the lessons from national experiences.

This report seeks to help build that trust by taking a detailed look at the climate change mitigation policies of OECD member countries and ten partner economies, building on recent OECD work and statistics. As emissions reductions are needed beyond the energy sector, the report also takes stock of actions in other areas of the economy such as agriculture, forestry, industry and waste, which can be significant sources of emissions in many countries.

It is encouraging to see that more and more countries are implementing policies and supporting research and development to reduce their emissions and promote low-carbon technologies in different sectors. However, the report concludes that greater efforts are needed to reach the targets and goals that have been announced and to achieve deep cuts in greenhouse gas emissions. The real challenge is not simply to meet an emissions reduction target in a given year, but to create credible pathways for each country towards a low-carbon future.

We are under no illusions regarding the difficulty of the task. For OECD member countries, it will not be easy to achieve the transition from carbon-intensive to low-emissions economies. For partner economies, the challenge is to avoid locking in emissions-intensive development paths. Our report shows that governments have made some important progress, but further effort is needed to deliver better climate change policies for better lives.

Angel Gurría
OECD Secretary General

Foreword

This report aims to increase transparency and improve understanding of different countries' situations by presenting trends and progress to date on climate change mitigation policies. It provides an overview of current policies to reduce greenhouse gas (GHG) emissions in the 34 OECD member countries and 10 partner economies (Brazil, People's Republic of China, Colombia, Costa Rica, Indonesia, India, Latvia, Lithuania, the Russian Federation and South Africa), as well as in the European Union. Together, these countries account for over 80% of global GHG emissions. It is a publication of the OECD Environment Directorate that was prepared as part of the Programme of Work and Budget 2015/16 of the Environment Policy Committee.

Action by countries to tackle climate change relates to national circumstances and emission profiles. This report presents an overview of mitigation targets and goals, together with a simple analysis of the changes in emission levels and improvements in carbon intensity needed by governments to achieve the targets and goals. The study examines recent developments in carbon pricing instruments such as energy and carbon taxation and emissions trading systems, as well as reforms to support for fossil fuels. It also reviews key domestic policy settings in the energy and non-energy sectors. The study focuses on climate change mitigation, while acknowledging that action on adaptation, finance, technology and capacity building is also a priority for many countries.

Multiple sources inform this report. It relies primarily on information gathered for the OECD Economic Surveys *in 2014 and 2015. Recognising that climate change is partly an economic issue, the OECD Economics Department (ECO) and the OECD Environment Directorate (ENV) have been working together to systematically address climate change issues in* OECD Economic Surveys *since 2014. The project also builds on efforts to mainstream green growth across the OECD's work programme following the launch of its* Green Growth Strategy *in 2011. In addition, it draws on other recent work by the OECD and the International Energy Agency (IEA), including on support for fossil fuels by the OECD Trade and Agriculture Directorate (TAD), on taxation of energy use by the OECD Centre for Tax Policy and Administration (CTP) and IEA and OECD statistics. National reports and greenhouse gas inventories submitted by Parties to the UN Framework Convention on Climate Change (UNFCCC) have also been useful sources.*

The report is accompanied by an online tool with country profiles containing more detailed information on the policies and progress of the countries studied. The online tool can be accessed at www.compareyourcountry.org/cop21.

Acknowledgements

This report is an output of the Environment Directorate, directed by Simon Upton. It was written and co-ordinated by Gregory Briner and Mikaela Rambali of the OECD Environment Directorate, under the guidance of Simon Buckle (Head of the Climate, Biodiversity and Water Division) and Robert Youngman (Head of the Climate, Finance and Investment team).

The authors would like to thank the following colleagues at the OECD and IEA Secretariats who provided valuable comments and expert review: Simon Upton, Anthony Cox, Simon Buckle, Robert Youngman, Geraldine Ang, Johanna Arlinghaus, Gérard Bonnis, Peter Borkey, Nils Axel Braathen, Laura Cozzi, Rob Dellink, Kurt van Dender, Olivier Durand-Lasserve, Jane Ellis, Florens Flues, Robert Ford, Justine Garrett, Guillaume Gruère, Michelle Harding, Christina Hood, Christopher Kaminker, Katia Karousakis, Takayoshi Kato, Osamu Kawanishi, Britta Labuhn, Lorcan Lyons, Virginie Marchal, Alen Maulitov, Sara Moarif, Luis Munuera, Yoko Nobuoka, Paul O'Brien, Ryan Parmenter, Andrew Prag, Roberta Quadrelli, Jehan Sauvage, Ronald Steenblik and Lola Vallejo. The authors would also like to thank Bernd Hackmann, Daniel Hooper, Kusum Lata and Katia Simeonova (UNFCCC) for their helpful comments. The authors would like to thank Mark Foss and Janine Treves for editorial assistance, Vincent Finat-Duclos for statistical editing and Sylvaine Herold for translation services.

Table of contents

Tables

Figures

Reader's guide

Countries included in the report

Country	Code	OECD member	Annex I Party	EU Member State
Australia	AUS	✓	✓	
Austria	AUT	✓	✓	✓
Belgium	BEL	✓	✓	✓
Brazil	BRA			
Canada	CAN	✓	✓	
Chile	CHL	✓		
China	CHN			
Colombia	COL			
Costa Rica	CRI			
Czech Republic	CZE	✓	✓	✓
Denmark	DNK	✓	✓	✓
Estonia	EST	✓	✓	✓
European Union	EU	✓	✓	
Finland	FIN	✓	✓	✓
France	FRA	✓	✓	✓
Germany	DEU	✓	✓	✓
Greece	GRC	✓	✓	✓
Hungary	HUN	✓	✓	✓
Iceland	ISL	✓	✓	
India	IND			
Indonesia	IDN			
Ireland	IRL	✓	✓	✓
Israel	ISR	✓		
Italy	ITA	✓	✓	✓
Japan	JPN	✓	✓	
Korea	KOR	✓		
Latvia	LVA		✓	✓
Lithuania	LTU		✓	✓
Luxembourg	LUX	✓	✓	✓
Mexico	MEX	✓		
Netherlands	NLD	✓	✓	✓
New Zealand	NZL	✓	✓	
Norway	NOR	✓	✓	
Poland	POL	✓	✓	✓
Portugal	PRT	✓	✓	✓
Russian Federation	RUS		✓	
Slovak Republic	SVK	✓	✓	✓
Slovenia	SVN	✓	✓	✓
South Africa	ZAF			
Spain	ESP	✓	✓	✓
Sweden	SWE	✓	✓	✓
Switzerland	CHE	✓	✓	
Turkey	TUR	✓	✓	
United Kingdom	GBR	✓	✓	✓
United States	USA	✓	✓	

Greenhouse gas emissions statistics

The following sources of greenhouse gas statistics are used in this document:

- **National GHG inventories**: *OECD Environment Statistics* (database) based on national inventory submissions to the United Nations Framework Convention on Climate Change (UNFCCC, CRF tables), and replies to the OECD State of the Environment Questionnaire. These statistics come from official submissions of GHG emissions data by Parties to the UNFCCC. Complete data sets including and excluding land use, land-use change and forestry (LULUCF) are available for Annex I Parties to the UNFCCC for 1990-2012 and partial data sets are available for non-Annex I Parties.
- **IEA statistics on CO_2 emissions**: IEA estimates of CO_2 emissions from fuel combustion are calculated using IEA energy data and the default methods and emission factors from the *Revised 1996 IPCC Guidelines for National Greenhouse Gas Inventories*. Time series are available for all 44 of the countries studied from 1990-2012.
- **IEA/EDGAR statistics on total GHG emissions:** This dataset combines IEA statistics on CO_2 from fossil fuel combustion with data for CO_2 from non-energy-related sources and gas flaring, and emissions of methane, nitrous oxide, hydrofluorocarbons, perfluorocarbons and sulphur hexafluoride from the Emissions Database for Global Atmospheric Research (EDGAR). The EDGAR database includes partial coverage of emissions from land use, land-use change and forestry (direct emissions from forest fires, emissions from decay of aboveground biomass that remains after logging and deforestation, emissions from peat fires and decay of drained peat soils). Statistics are available for all 44 countries studied for the years 1990, 2000, 2005 and 2010.

Energy statistics

This report generally follows the IPCC definition of the energy sector used in the context of national GHG inventories. In this context, the "energy" category includes fossil fuel combustion from energy industries, transport, manufacturing and construction, and other sectors, as well as fugitive emissions from fuels. This comprises energy extraction, conversion, storage, transmission and distribution processes that deliver final energy to the end-use sectors (industry, transport and building, as well as agriculture and forestry). Given the common distinction made by the policy-making community between transport and other energy sub-sectors, transport is discussed separately from other energy sub-sectors in this report.

Cut-off date

Climate change mitigation policies are rapidly evolving in many countries. The cut-off date for inclusion of policy developments in this report was August 2015.

Abbreviations and acronyms

BAU	Business-as-usual
CCS	Carbon Capture and Storage
CH_4	Methane
CO_2	Carbon dioxide
CO_2e	Carbon dioxide equivalent
COP	Conference of the Parties
ETS	Emissions Trading System
FAO	Food and Agriculture Organization of the United Nations
GDP	Gross domestic product
GHG	Greenhouse gases
$GtCO_2e$	Gigatonnes of carbon dioxide equivalent
HFC	Hydrofluorocarbon
IEA	International Energy Agency
INDC	Intended Nationally Determined Contribution
IPCC	Intergovernmental Panel on Climate Change
LULUCF	Land use, land-use change and forestry
LPG	Liquefied petroleum gas
$MtCO_2e$	Million tonnes of CO_2 equivalent
Mtoe	Million tonnes of oil equivalent
N_2O	Nitrous oxide
OECD	Organisation for Economic Co-operation and Development
PFC	Perfluorocarbon
PPP	Purchasing Power Parity
RD&D	Research, development and demonstration
SF_6	Sulphur hexafluoride
TPES	Total primary energy supply
tCO_2e	Tonnes of carbon dioxide equivalent
UNEP	United Nations Environment Programme
UNFCCC	United Nations Framework Convention on Climate Change
USD	United States Dollars

Climate Change Mitigation
Policies and Progress
© OECD 2015

Executive summary

Tackling climate change is a difficult political challenge requiring a high level of trust and co-operation between countries. Global greenhouse gas (GHG) emissions need to be 40-70% below 2010 levels by 2050 and near zero or negative by 2100 to hold the rise in global average temperature to below 2 °C. If current trends continue, there is a high probability of significantly greater temperature rises, increasing the risk of severe and irreversible impacts on ecosystems, significant disruptions to agricultural systems and impacts on human health in this century and beyond.

This report presents trends and progress on climate change mitigation policies in the 34 OECD member countries, the European Union and 10 partner economies (Brazil, the People's Republic of China, Colombia, Costa Rica, Indonesia, India, Latvia, Lithuania, the Russian Federation and South Africa). It is intended to increase transparency and improve understanding of mitigation goals and the extent to which carbon pricing instruments and other policies to address GHG emissions have been implemented across different economic sectors.

The following key developments relating to climate change mitigation policies are identified:

- **Aggregate GHG emissions from the countries studied have been increasing since the 1990s, although GHG emissions per unit of gross domestic product (GDP) have decreased in nearly all cases.** In several cases, emissions declined in recent years in the wake of the financial crisis, but have since rebounded due to increased economic activity or changes in nuclear energy policy following the Fukushima nuclear accident. While some countries have reduced their emissions, more ambition is needed by all, in line with the principles of the UN Framework Convention on Climate Change (UNFCCC), to avoid dangerous human-caused climate change.
- **While use of low-carbon energy sources is increasing, most countries still rely on fossil fuels to power their economies and continue to support the production and consumption of fossil fuels.** In particular, coal – the most carbon-intensive fuel – still accounted for 45% of electricity generation in the countries studied in 2012. Although several countries have made progress on reforming subsidies for fossil-fuel consumption, many countries continue to support fossil-fuel production and consumption.
- **Taxes on energy are gradually being re-oriented to reflect the carbon content of fuels and an increasing number of jurisdictions are using carbon taxes to explicitly price CO_2 emissions.** However, the share of total emissions covered by energy and carbon taxes remains low and tax rates to date have been insufficient to spur technological change and significantly alter consumer behaviour. Carbon taxes are implemented or planned at the national or sub-national level in 15 of the countries studied.

- **An increasing number of international, national and sub-national jurisdictions are implementing emissions trading systems (ETSs), but allowance prices are low.** ETSs have been established in the European Union and at the national level in Korea, New Zealand and Switzerland. China has launched pilot ETSs in seven cities and provinces and is planning to launch an economy-wide system. Sub-national ETSs have also been implemented in California and nine north-eastern US states, Quebec in Canada, and Tokyo and Saitama in Japan.
- **Several of the countries studied have recently reformed their renewable energy support policies, with decreased use of feed-in tariffs and increased use of feed-in premiums and competitive bidding processes.** Emission standards for power plants, fuel economy standards for vehicles and energy efficiency standards for buildings are also widely used. Other objectives such as improving energy security, air quality and human health can be drivers for such policies.
- **Public spending on energy-related RD&D as a share of GDP remains low, although the share of energy-related RD&D spending allocated to low-carbon energy technologies such as energy storage, smart grids, advanced fuels and vehicles, and carbon capture and storage (CCS) is rising.** In 2012, 22 OECD member countries collectively spent around USD 13 billion on public energy-related RD&D, mainly for renewable energy sources, energy efficiency and nuclear energy. CCS accounts for over half of public RD&D spending on fossil fuels in certain countries. The private sector is also an important source of energy-related RD&D spending.
- **While most of the countries studied have taken limited action to date to reduce emissions from agriculture, some countries have made significant progress to reduce deforestation and are addressing GHG emissions from other non-energy sectors.** Agriculture, deforestation, industrial processes and waste are significant sources of GHG emissions in some countries. The emissions intensity of the agriculture sector has decreased in many countries since 1990. However, implementing mitigation policies in this sector has proven challenging partly due to the limited availability of low-cost agricultural mitigation technologies in many regions. Significant progress has been made on reducing deforestation rates in some countries (e.g. Brazil), albeit from a high starting point. Mixes of economic instruments, regulations and information programmes are being used to reduce GHG emissions from the industry and waste sectors.

Almost all of the countries studied have taken on mitigation targets or goals for 2020 in the context of the UNFCCC or the Kyoto Protocol, with the nature and ambition of these goals reflecting national circumstances. Many have also announced intended nationally determined contributions (INDCs) for the post-2020 period. At the national level, the United Kingdom has established a legally-binding, long-term mitigation target together with short-term carbon budgets, and similar approaches have been established or are under consideration in Denmark, Finland, France and Norway. Many countries have also set national targets for relevant indicators such as GHG emissions, renewable energy, energy efficiency and forest cover.

Even if the INDCs and national targets announced to date are fully achieved, the remaining global carbon budget (consistent with a below 2 °C world) will be exhausted by around 2040 unless stronger action is taken. Although most of the countries studied are making some progress towards meeting their mitigation targets and goals, many are on a trajectory that is likely to fall short in the absence of a significant acceleration in annual emission reduction rates.

Climate Change Mitigation
Policies and Progress

Chapter 1

The state of play of climate change mitigation policies

This chapter provides an overview of the different national circumstances and emissions profiles that are needed to put climate mitigation policy responses into context. It also includes a summary of international mitigation targets, and goals and information on domestic climate policy settings. The policies described include national climate change plans and domestic targets, carbon and energy taxation, emissions trading systems, support for fossil fuels, innovation and research, development and demonstration (RD&D), renewable energy support policies, regulatory standards and policies to reduce emissions and enhance sinks in other sectors. This chapter serves as an extended summary of the main messages of the report.

The statistical data for Israel are supplied by and under the responsibility of the relevant Israeli authorities. The use of such data by the OECD is without prejudice to the status of the Golan Heights, East Jerusalem and Israeli settlements in the West Bank under the terms of international law.

Deep and sustained cuts in greenhouse gas (GHG) emissions are needed. Governments have agreed in the context of the United Nations Framework Convention on Climate Change (UNFCCC) to hold the increase in global average temperature to below 2 °C above pre-industrial levels in order to avoid the worst impacts of climate change (UNFCCC, 2010). Given the long atmospheric lifetime of carbon dioxide (CO_2), any target to stabilise the global average temperature implies that zero net emissions will need to be achieved. Achieving the 2 °C objective will require global emission reductions of 40-70% from 2010 levels by 2050 and near-zero emissions of CO_2 and other long-lived greenhouse gases by the end of the century (IPCC, 2014). The timing of the global emissions peak and decline changes the probability distribution of the global average temperature increase; later peaking increases the likelihood of higher temperature rises. Enhanced carbon sinks and negative-emission technologies may be needed to achieve this objective. The scale of the challenge demands enhanced action and co-ordination between all actors, including national and sub-national governments, the private sector and civil society.

Different emissions profiles and starting points for countries

Any analysis of climate change mitigation policies needs to take into account national circumstances and the fact that all countries are at different starting points. Gross domestic product (GDP) per capita and GHG emissions per capita vary widely across the countries studied (Figure 1.1). Annual GDP per capita ranges from USD 2 300 in India to USD 69 600 in Luxembourg, while annual GHG emissions per capita range from 1.5 tCO_2e in India to 24 tCO_2e in Australia. While GHG emissions have increased in absolute terms in many countries, GHG emissions per unit of GDP decreased in nearly all of the countries studied between 1990-2012 – albeit at different rates and from different starting points (Figure 1.2). These GHG emissions figures are calculated on a production basis; the picture changes slightly if CO_2 embodied in international trade is taken into account (Box 1.1).

All countries need to cut emissions in all key sectors to meet the climate challenge, while taking into account national circumstances and in accordance with the principles of the UNFCCC. Energy, including power generation and transport, accounts for over 70% of total GHG emissions for most OECD member countries (Figure 1.4). Policies addressing GHG emissions from energy are therefore the main focus of this report.

GHG emissions from land use, land-use change and forestry (LULUCF), agriculture, industry and waste also make significant contributions in some countries (Figure 1.4). In this report, the term LULUCF refers to emissions and removals from forest and other land uses. In countries with extensive forests and low deforestation levels, LULUCF can be a net sink of GHG emissions. Policies addressing CO_2 and non-CO_2 gases in other sectors (including CO_2 removals by forests and other land sinks) are therefore also outlined in the report.

Figure 1.1. **GHG per capita and GDP per capita**

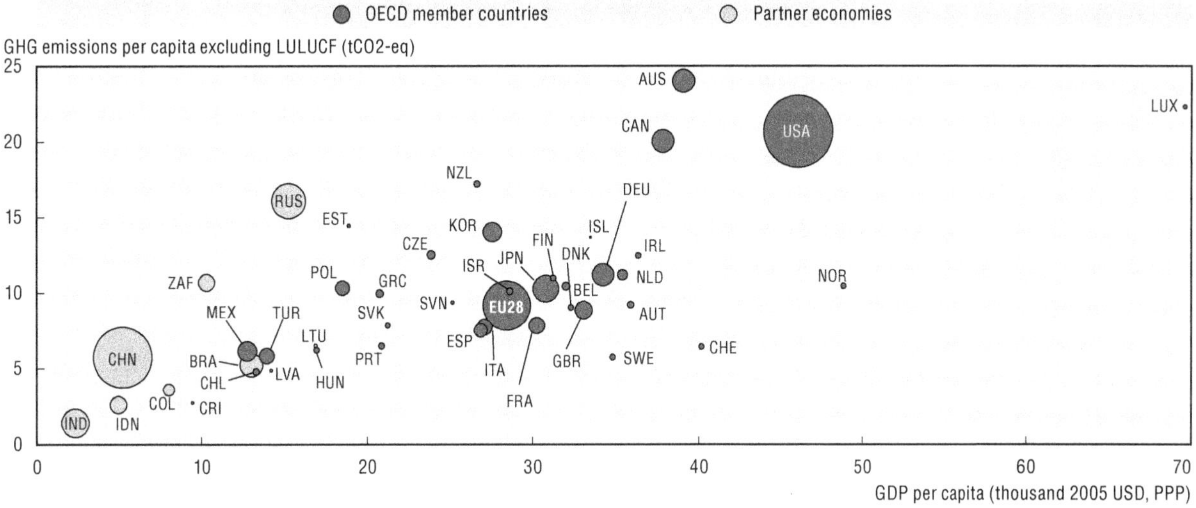

1. Values for 2012 except Chile (2006), China (2005), Colombia (2004), Costa Rica (2005), India (2000), Indonesia (2000), Israel (2011), Korea (2011), Mexico (2010) and South Africa (2010).
2. Bubble size is proportional to total GHG emissions.

Source: GDP and population statistics from IEA (2015c), "Indicators for CO_2 emissions", *IEA CO_2 Emissions from Fuel Combustion Statistics* (database), *http://dx.doi.org/10.1787/data-00433-en* (accessed 21 July 2015); OECD (2015g), *OECD Environment Statistics* (database) (accessed 07 July 2015); UNFCCC (2015), GHG inventory data (accessed 07 July 2015).

StatLink http://dx.doi.org/10.1787/888933272394

Figure 1.2. **Trends in greenhouse gas emissions per unit of GDP**

Percentage change 1990-2012*

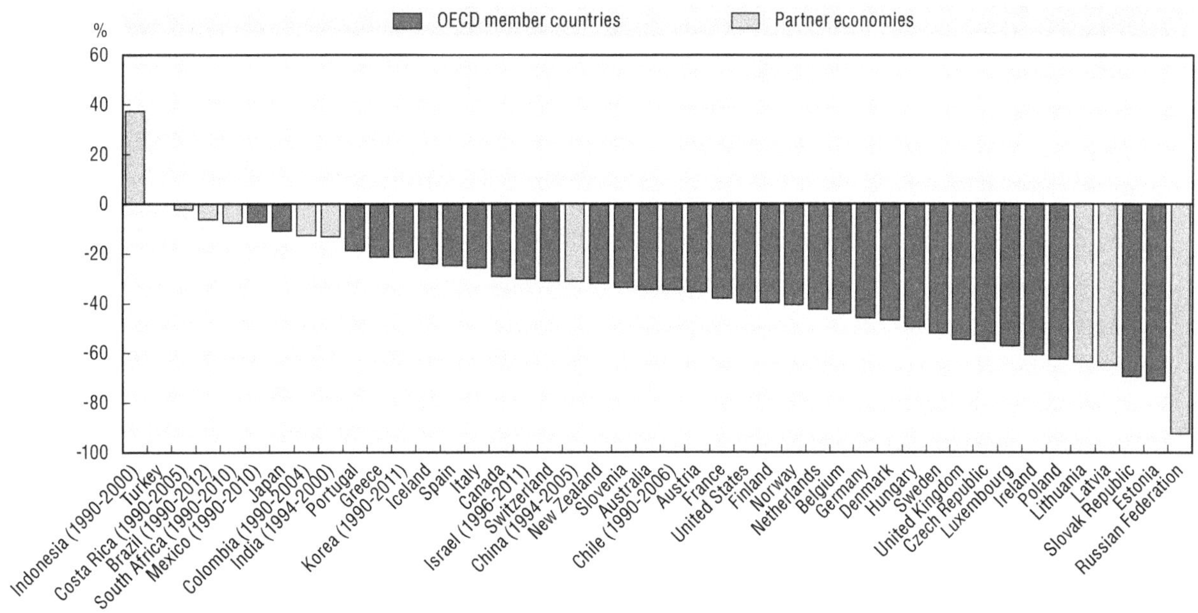

* 2012 or latest year available.
Greenhouse gas emissions per unit of GDP (USD 2005 PPP).

Source: GDP statistics from IEA (2015c), "Indicators for CO_2 emissions", *IEA CO_2 Emissions from Fuel Combustion Statistics* (database), *http://dx.doi.org/10.1787/data-00433-en* (accessed 21 July 2015); GHG statistics from OECD (2015g), *OECD Environment Statistics* (database) (accessed 07 July 2015), and UNFCCC (2015), GHG inventory data (accessed 07 July 2015).

StatLink http://dx.doi.org/10.1787/888933272404

Box 1.1 **Embodied CO_2 emissions**

GHG inventories under the UNFCCC are calculated on a production basis, i.e. taking into account emissions released within the geographic boundaries of each country. In many cases, however, GHG emissions result from the production of goods that are exported and consumed in a different country. The relative magnitudes of countries' CO_2 emissions can therefore look different when the effect of embodied carbon in imports and exports is taken into account (Figure 1.3).

The OECD has estimated embodied CO_2 in international trade for 61 countries and the rest of the world, based on its Inter-Country Input-Output (ICIO) database, the IEA's CO_2 emissions from fuel combustion statistics and other industry data. The People's Republic of China (hereafter "China") had the highest absolute emissions in 2011 from both consumption and production perspectives. While Chinese per capita emissions have doubled since 1995, consumption-based per capita emissions in the United States are still almost four times larger. For some countries, up to 75% of the emissions embodied in the final goods and services consumed are emitted elsewhere in the world, while this number is less than 10% for others (OECD, 2015a).

Figure 1.3. **Embodied per capita CO_2 emissions from fossil fuel combustion**

Tonnes of CO_2 equivalent in 2011

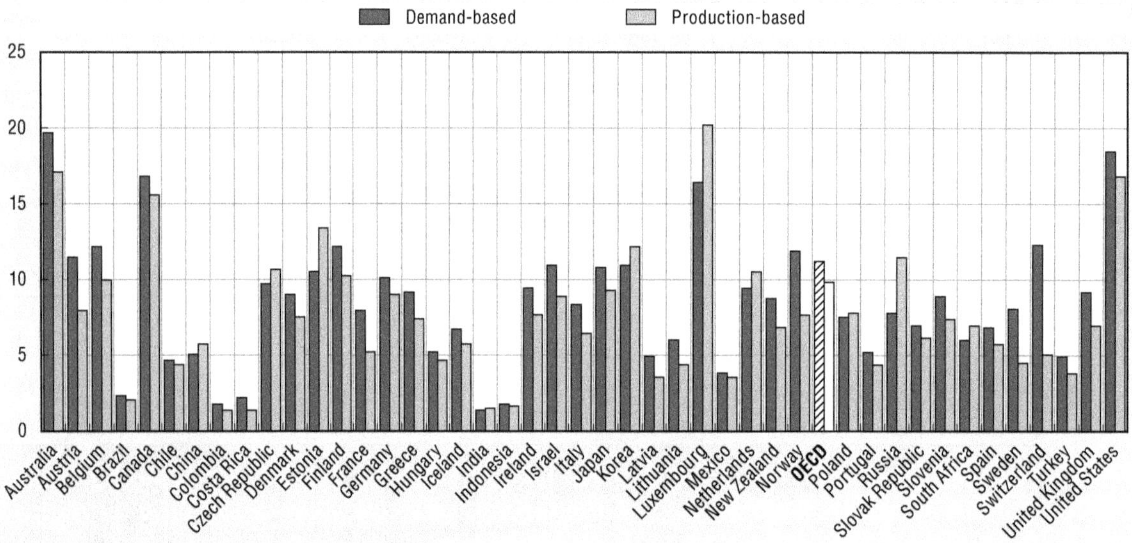

Source: OECD (2015a), *OECD Embodied Carbon Dioxide Emissions*, edition 2015, *www.oecd.org/sti/inputoutput/co2* (accessed 20 July 2015).

StatLink http://dx.doi.org/10.1787/888933272412

Aggregate GHG emissions from the countries studied have continued to grow over the past decade. In some cases, such as the European Union and the United States, CO_2 emissions from energy use have declined in recent years due to switching fuel to less carbon-intensive sources (e.g. from coal to natural gas), increased use of renewable energy technologies and improved energy efficiency, and more recently, the impact of the global economic crisis. However, some emission reduction rates are now faltering due to (i) resumption of activity after the global economic crisis; (ii) replacement of nuclear power with more CO_2-intensive forms of power generation in a number of countries following the accident at Fukushima Daiichi nuclear power station in Japan; (iii) growth in demand for transport; and (iv) increased use of coal without carbon capture and storage (CCS) for power generation. Rapidly scaled-up deployment of CCS will be needed to meet climate policy objectives if coal continues to be used for power generation.

Figure 1.4. **Greenhouse gas emissions by sector**

Percentage of total GHG emissions excluding LULUCF in 2012

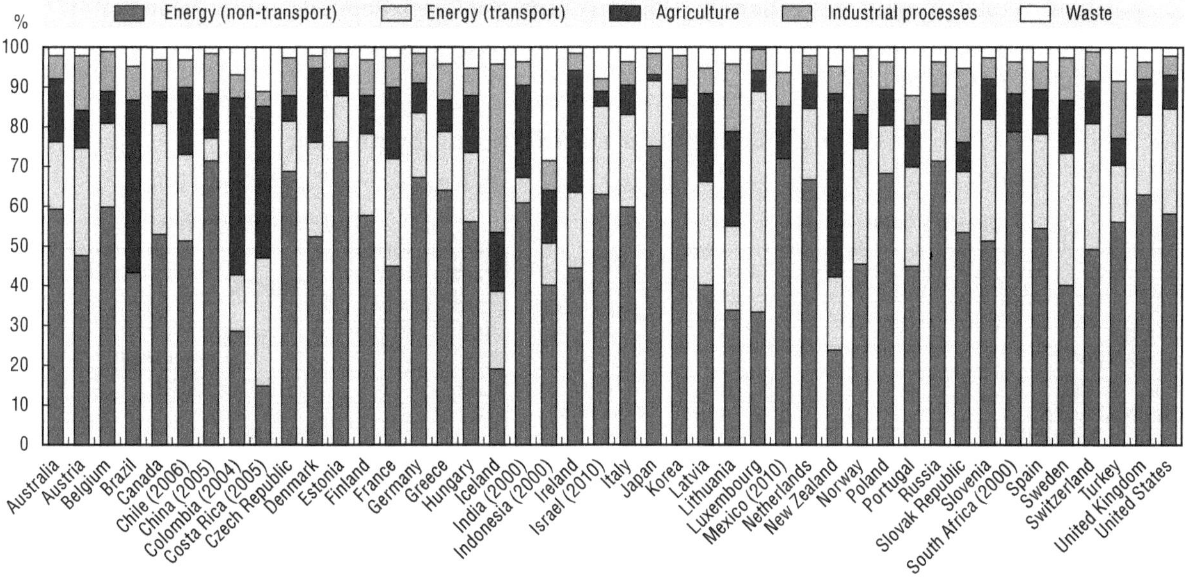

1. 2012 or latest year available. Excludes LULUCF.
2. Energy (non-transport) includes energy industries, manufacturing and construction, and other sectors; fugitive emissions from fuels.
3. Due to data limitations, emissions from transport are not separated from energy for Korea, Mexico and South Africa.

Source: Data by sector from OECD (2015g) OECD Environment Statistics (database), "Greenhouse gas emissions by source", (accessed 07 July 2015); UNFCCC (2015), GHG inventory data (accessed 07 July 2015).

StatLink http://dx.doi.org/10.1787/888933272428

In the energy sector, CO_2 emissions can be reduced by cutting energy consumption and decreasing the carbon intensity of the energy mix. The emission abatement potential and costs typically vary between countries, as well as between sub-sectors (e.g. power generation, heat, transport). Most countries continue to rely heavily on fossil fuels to power their economies, although many promote low-carbon energy sources such as renewables and nuclear power. Several countries have shifted from coal towards gas since 1990. This shift lowers GHG emissions in the near term, since combustion of gas releases lower CO_2 emissions per unit of energy than coal (in the absence of carbon capture and storage technology); however, it still risks locking in an energy system based on fossil fuels over the longer term. Renewable energy sources now supply the greatest share of energy in Costa Rica, Finland, Iceland, Latvia, New Zealand, Norway and Sweden (Figure 1.5). Nuclear power provides the largest share in France and Belgium.

The energy mixes of countries are shaped by many different and often inter-linked factors. These include political and economic priorities, geography and natural resource endowments, natural disasters, public opinion and energy prices. Some factors have a positive impact on GHG emissions, while others have a negative impact. For example, using domestic energy sources can be cheaper than importing energy and can increase energy security. Natural endowments can therefore significantly influence the energy mix. This can be observed in the high share of domestic oil shale in the electricity generation mix of **Estonia** (OECD, 2015b), the high share of geothermal and other renewable energy sources in **Iceland** and **New Zealand**, the high share of hydro power in the electricity generation mix of **Brazil**, **Canada** and **Norway** and the high share of coal in the electricity generation mix of **China**, **India**, **Poland** and **South Africa**. Coal still accounts for the greatest share of total primary energy supply (TPES) in **China**, the **Czech Republic**, **India**, **Poland** and **South Africa**, and is still being used to help meet incremental demand in several partner economies (Figure 1.5).

Figure 1.5. **Energy and electricity mix by product**

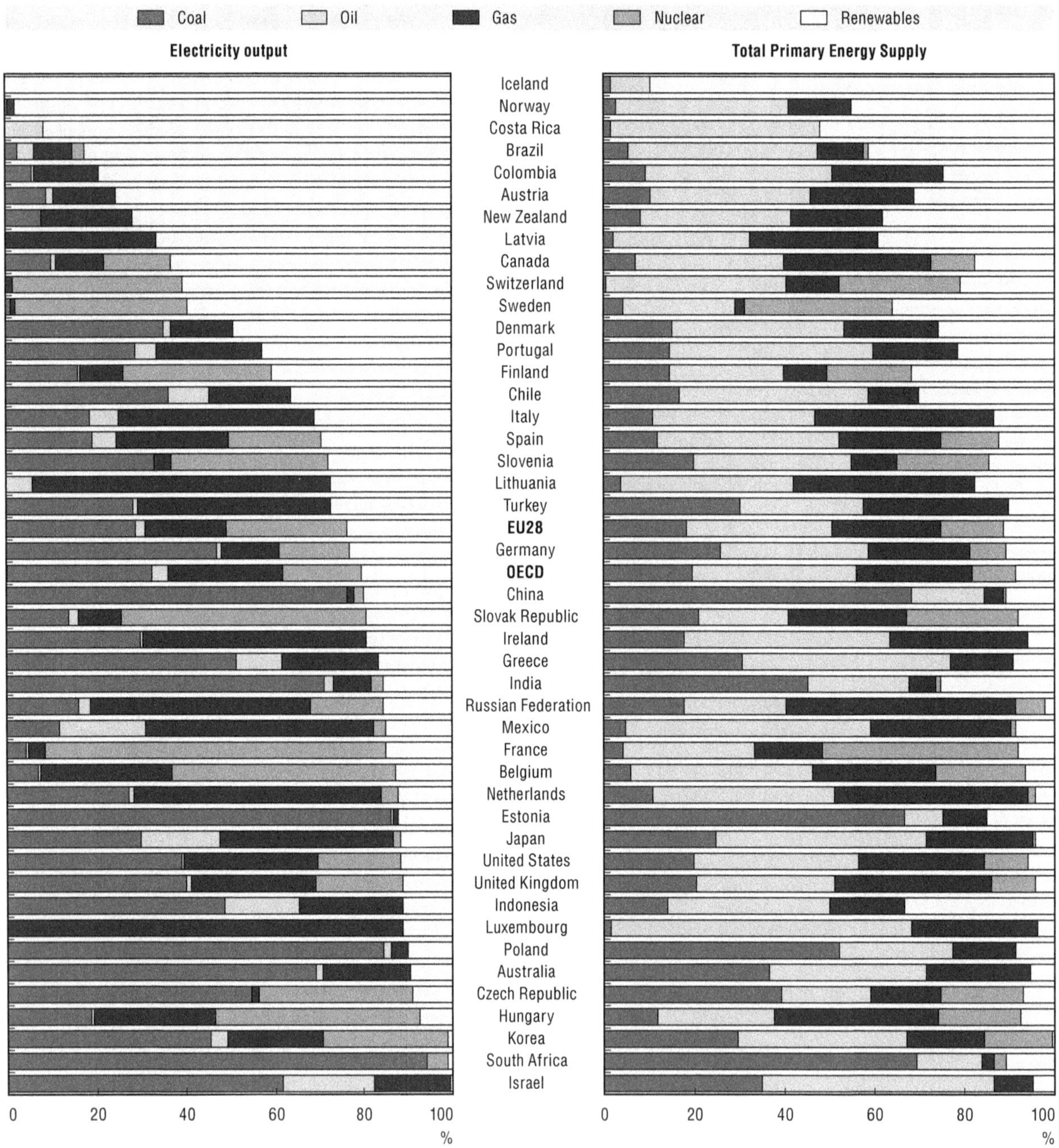

Source: IEA (2015b), "World Energy Balances", IEA World Energy Statistics and Balances (database), *http://dx.doi.org/10.1787/data-00512-en* (accessed 24 April 2015).

StatLink http://dx.doi.org/10.1787/888933272437

Recent developments in the nuclear power industry could have significant long-lasting implications for GHG emissions. Following the Great East Japan Earthquake in 2011, **Japan** is reviewing the safety of all 43 nuclear reactors not set for decommissioning (JAIF, 2015). The decrease in nuclear generation capacity has been largely replaced by power plants fired by fossil fuels, which have led to higher CO_2 emissions in Japan (OECD, 2015c). Subsequently, **Germany** decided to phase out the use of nuclear power by 2022, a decision that has led to increased consumption of coal (as well as renewable energy). Similarly, **Belgium and Switzerland** have decided to phase out nuclear power.

In the **United States**, unconventional natural gas extracted by hydraulic fracturing ("fracking") has displaced coal in the electricity generation mix and led to a decrease in GHG emissions (OECD, 2014a). However, concerns about potentially harmful environmental and health impacts have led to bans on fracking in some US states and municipalities (e.g. New York State, Maryland and Vermont) as well as **France**, where a 2011 decision to ban fracking has since been upheld (Conseil Constitutionnel, 2013). By contrast, production of unconventional crude oil and bitumen from oil sands in **Canada** increased by nearly 450% between 1990-2012, leading to an 82% increase in GHG emissions from the fossil fuel extractive industries sector over this period (Government of Canada, 2014).

Trends in energy demand also vary significantly between countries. Many factors affect energy demand, including economic growth, climate, population density, the efficiency of energy industries and the efficiency of transmission and distribution systems (Box 1.2). Energy supply is increasing fastest in rapidly growing economies such as **Chile, China, India and Turkey** (Figure 1.6). Between 1990-2012, the collective total primary energy supply (TPES) of the 10 partner economies more than doubled on average, compared with 16% growth for OECD member countries. Thus, increasing energy efficiency remains critical for decreasing global GHG emissions, particularly in emerging economies with major potential for abatement.

Box 1.2. **Energy efficiency in power transmission and distribution**

Higher energy efficiency in transmission and distribution (T&D) can mitigate emissions. Electrical losses result from cable or line losses coupled with transformer losses; these vary according to the geographical layout and nature of the power system. Losses can come from ageing equipment, network congestion or extreme peak load demand, and result in increased GHG emissions. The International Energy Agency (IEA) estimates that around USD 1 787 billion of cumulative investment in transmission and distribution is needed globally in 2014-35 to satisfy increasing demand for energy services (IEA, 2014a). In India, for instance, while energy demand has increased rapidly with economic growth, transmission and distribution losses (including technical losses and theft) hamper energy supply. The electricity system in India suffers from frequent blackouts resulting from an inefficient national transmission grid. It is estimated that more than 20% of Indian electricity output in 2011 was lost in T&D (OECD, 2014b).

Source: IEA (2014a) *World Energy Investment Outlook 2014*; OECD (2014b) *Economic Survey of India*.

Other benefits besides climate change mitigation can be strong drivers of domestic policy making. Renewable energy and other clean energy technologies can contribute to other policy objectives, including improving energy security, air quality, health improvements, flood protection, and biodiversity and ecosystem conservation. For many partner economies, reducing poverty, increasing energy access, increasing food and water

Figure 1.6. **Trends in total primary energy supply**

Percentage change, 1990-2012

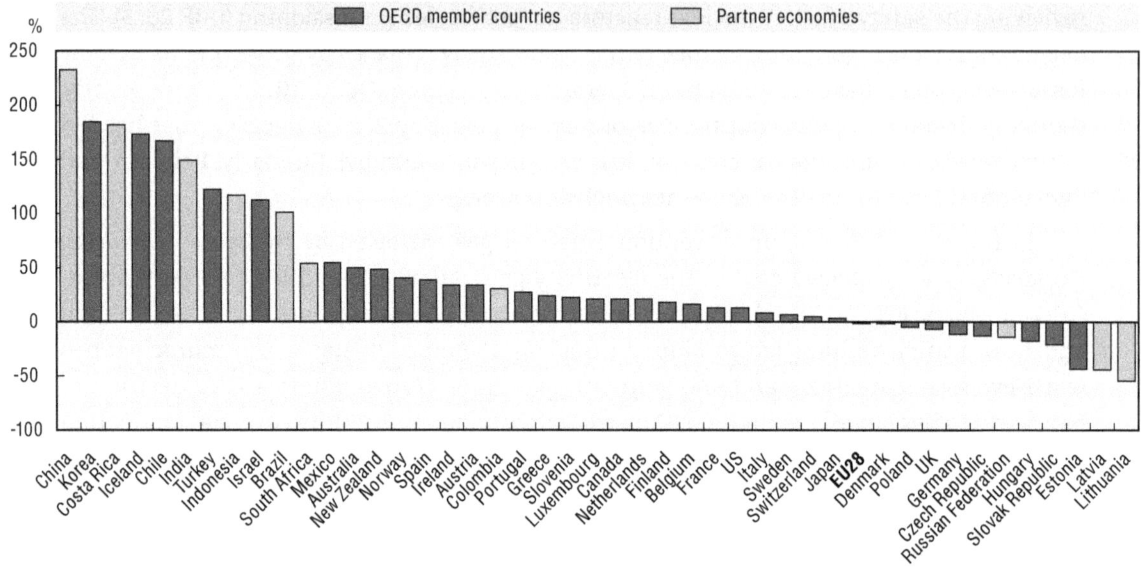

Source: IEA (2015b), "World Energy Balances", IEA World Energy Statistics and Balances (database), http://dx.doi.org/10.1787/data-00512-en (accessed 24 April 2015).

StatLink http://dx.doi.org/10.1787/888933272445

security, and promoting rural economic development still remain high priorities. Millions of people still lack access to electricity in partner economies (IEA, 2014b). Many of these countries still rely heavily on traditional use of solid biomass for cooking, and the amount of energy used per capita is very low. Policies are needed that can help improve energy access and reduce poverty, while avoiding development pathways that lock in dependence on traditional carbon-intensive energy technologies. For instance, decentralised renewable energy systems can increase access to energy in remote off-grid areas.

Policies such as increasing use of clean energy and improving energy efficiency can have multiple benefits. India's National Action Plan on Climate Change, for instance, emphasises that tackling GHG emissions through supporting renewable energy with its National Solar Mission can also provide higher energy security (Government of India, 2008). China highlights the economic gains from mitigation through energy efficiency, as well as health improvement in its 12th Five-Year Plan (2011-15) (CBI China, 2012). Energy efficiency can reduce energy demand and associated costs, as well as contribute to achieving other objectives such as making indoor environments healthier (IEA, 2014c). Better air quality can also reduce health costs. The OECD has estimated the annual cost of air pollution to OECD societies, China and India to be USD 3.5 trillion in terms of the value of lives lost and ill health (OECD, 2014c).

Multilateral action to address climate change

The main channel for multilateral action on climate change is the United Nations Framework Convention on Climate Change (UNFCCC). This convention, signed in 1992, has near global participation and its ultimate objective is "stabilisation of greenhouse gas concentrations in the atmosphere at a level that would prevent dangerous anthropogenic interference with the climate system" (UNFCCC, 1992). In 1997, Parties adopted the Kyoto Protocol to the UNFCCC, which sets binding emission reduction commitments for most

Annex I Parties (industrialised countries that were members of the OECD in 1992, plus a number of countries in transition to a market economy). The first commitment period of the Kyoto Protocol was 2008-12 and the second is 2013-20. Fewer Annex I Parties chose to participate in the second commitment period, and the share of global emissions covered by the Kyoto Protocol decreased from around 22% in the first commitment period to around 13% in the second commitment period.

In 2010, countries agreed to work together to limit the increase in global average temperature to below 2 °C above pre-industrial levels. Governments were invited to pledge mitigation targets and actions for 2020 under the convention. This was significant as it was the first time that developing countries had pledged to undertake specific mitigation actions under the UNFCCC. All Annex I Parties except Turkey pledged emission reduction targets for 2020 relative to a base year. Non-Annex I Parties expressed their mitigation actions using a range of different metrics. These included reductions in emission intensity (**China, India**), reductions in GHG emissions relative to business-as-usual (BAU) baselines (**Brazil, Indonesia, Israel, Korea, Mexico, South Africa**) and achievement of carbon neutrality (Costa Rica), as well as other sector-specific targets for increasing the use of renewable energy sources, increasing the use of biofuels and/or enhancing forest cover and stock volume (**China, Colombia, Israel**).

A new multilateral framework for tackling climate change is being negotiated. Parties to the UNFCCC are working towards a new climate deal at COP 21 in Paris at the end of 2015. The expected agreement, to be implemented from 2020, should cover several topics, including mitigation, adaptation, finance, technology, capacity building, transparency and implementation. Climate change is part of a wider set of inter-linked environmental and developmental challenges. To reflect this, links are being created between the UNFCCC and other multilateral processes such as the post-2015 Development Agenda.

In the lead up to COP 21 in Paris, countries are announcing their intended mitigation targets and goals for the post-2020 period (known as "intended nationally determined contributions", or INDCs). Similarly to the 2020 mitigation pledges, these INDCs are being expressed using a variety of metrics. Countries are also using different time scales and taking different approaches to the use of credits from market mechanisms, as well as treatment of LULUCF. This tailored approach enables countries to express targets and goals that best reflect their national circumstances. However, it presents challenges in terms of comparability of INDCs and estimating the future global emissions pathway in relation to the 2 °C goal. Countries have been encouraged to submit information alongside their INDCs to facilitate understanding and transparency, which is important for building trust and increasing ambition over time. While the INDCs are mainly focused on mitigation, some also include an adaptation component.

This report finds that while many countries are making some progress towards meeting their mitigation targets and goals, many are on a trajectory that is likely to fall short in the absence of a significant acceleration in annual emission reduction rates. This can be achieved through further decoupling of GHG emissions growth from GDP growth in the coming years. The IEA's *World Energy Outlook Special Report on Energy and Climate* (IEA, 2015) provides an in-depth analysis of INDCs with a focus on the energy sector. The report estimates that if current trends continue, the world's remaining carbon budget (consistent with a 50% chance of keeping the rise in temperature below 2 °C) will be consumed by around 2040, and the INDCs submitted to date will only delay the complete exhaustion of the global carbon budget by eight months.

For countries with absolute emission reduction targets for 2020, the average annual emission reduction rates excluding LULUCF needed from 2012 to meet the targets range from +2.8% to –7.1% per year. Positive rates of change of emissions indicate that a net increase in emissions is possible under the target or goal, while negative rates imply that emission reductions are required. For emission reduction targets beyond 2020, the range is from –0.5% to –4.6% per year. Different reduction rates are obtained if LULUCF is included. For countries with emission intensity goals, annual reductions in emission intensity of between 0.3% and 3.9% are needed to meet the 2020 goals. For countries with mitigation goals relative to BAU baselines, significant deviations from BAU annual emission growth rates are needed. Some countries with mitigation goals relative to BAU baselines have not published these baselines, which makes it difficult to assess progress. Further, recent GHG statistics are unavailable or incomplete for many partner economies.

National and sub-national action to tackle climate change

All of the countries studied have set up institutional frameworks to address climate change. In many cases, a key challenge is to improve co-operation between national ministries and government departments, as well as between national and sub-national levels of government. Some countries have enhanced government co-ordination on climate policy by establishing inter-ministerial institutions, such as the **Indian** Prime Minister's Council on Climate Change, **Mexico's** Commission on Climate Change and **Japan's** Global Warming Prevention Headquarters. Other countries have set up independent bodies to provide information, guidance and policy recommendations, such as the **UK's** Committee on Climate Change established in 2008. Finance and economic ministries typically play a key role in integrating climate change and other green growth objectives into economic policy making and development planning (OECD, 2015d). Climate policy can be more effective if all government ministries identify key misalignments with low-carbon transition in their portfolios (OECD, IEA, NEA, ITF, 2015; Box 1.3).

Many countries have established national climate change plans and set domestic targets to complement their international objectives. Some countries are aspiring to reach carbon neutrality, such as **Costa Rica** (starting year 2021), **Sweden** (by 2050) and **Norway** (conditional by 2030 and unconditional by 2050). Norway's 2030 pledge is conditional upon commitments of other developed nations, and effectively means a commitment to reduce emissions abroad equivalent to Norwegian emissions in 2030. The **United Kingdom** has put in place a system of legally binding short-term carbon budgets that restrict the total amount of GHG emissions it can emit over five years. These are aligned with the country's long-term objective to reduce GHG emissions by 80% by 2050. **Denmark** has also passed a Climate Change Act in 2015 that establishes a similar system of short-term emissions budgets linked to long-term emissions pathways. **Finland's** Parliament approved a Climate Change Act in 2015 that establishes an emission reduction target of 80% from 1990 levels by 2050, with long-term mitigation action plans to be drawn up every ten years (Finland Ministry of the Environment, 2015). Countries have also set sector-specific targets. For instance, over 160 countries have set renewable energy or energy efficiency targets (REN21, 2015; IRENA, 2015).

Important mitigation actions are also being taken at the sub-national level and by non-state actors, including civil society, local authorities and private companies. The French presidency has highlighted enhanced co-operation between these actors (known as the "Solutions Agenda", Box 1.4) as one of its four pillars for COP 21 (Government of

Box 1.3. **Aligning policies for the transition to a low-carbon economy**

At the 2014 OECD Ministerial Council Meeting, ministers and representatives of OECD member countries and the European Union invited the OECD to work with the International Energy Agency (IEA), the International Transport Forum (ITF) and the Nuclear Energy Agency (NEA) "to continue to support the UNFCCC negotiations and to examine how to better align policies across different areas for a successful economic transition of all countries to sustainable low-carbon and climate-resilient economies and report to the 2015 OECD Ministerial Council Meeting." The OECD report on aligning policies for a low-carbon economy (OECD, IEA, ITF, NEA, 2015) responds to that request by identifying where existing policy and regulatory frameworks are at odds with core climate policy, i.e. where existing policies may make climate policy less effective than it could be otherwise. It reflects the initial diagnosis on where and how existing policy and regulatory frameworks may not be aligned with a low-carbon economy.

The areas examined include policies related to the economy, fiscal and financial issues, competition, employment, society, environment, energy, investment, trade, development co-operation, innovation, agriculture and sustainable food production, transport, and regional and urban issues. The report concludes that climate policy can be more effective if all government ministries identify important misalignments with low carbon transition in their respective portfolios. An ambitious climate action plan will therefore need new approaches to policy making across government. Beyond the national level, better alignment of policies across countries could also boost effectiveness and alleviate concerns about potential distortions of competition.

Source: OECD, IEA, ITF, NEA (2015), *Aligning Policies for a Low-carbon Economy*, OECD Publishing, Paris, *http://dx.doi.org/10.1787/9789264233294-en*.

Box 1.4. **The Solutions Agenda**

Following the Climate Summit in New York in 2014 and launched during COP 20 in Peru, the Solutions Agenda was developed by the Peruvian and French COP presidencies. The Solutions Agenda refers to climate action by sub-national authorities, private companies and economic sectors. It emphasises co-operative initiatives from governmental and non-governmental actors (businesses, local governments, international organisations, NGOs, indigenous peoples, etc.), as well as individual commitments by local and regional governments and businesses. The Solutions Agenda is the fourth pillar identified by the French presidency for COP 21 (France Ministry of Foreign Affairs and International Development, 2015; Government of France, 2015).

Examples of actions that could fit into the Solutions Agenda include Japan's Joint Crediting Mechanism (JCM) and city-to-city partnerships between Japanese and other local governments in Asia; the pledge by more than 130 governments, companies, civil society and indigenous peoples' groups to halve deforestation by 2020, and to end it by 2030; and the compact between mayors of more than 2 000 cities around the world to curb GHG emissions by 454 $MtCO_2e$ by 2020 (Government of France, 2015).

France, 2015). The United Nations Environment Programme (UNEP) has identified over 180 such co-operative initiatives; it estimates that existing climate initiatives involving cities, companies and sectors could save 2.9 $GtCO_2e$ by 2020, with a range of 2.5-3.3 $GtCO_2e$ corrected for overlap between initiatives (UNEP, 2015).

In federal states, mitigation action is often undertaken at the sub-national level with co-ordination at the central level. For example, **China** uses its system of five-year national development plans to establish sub-national plans with their own specific policies and targets. China's 12th Five-Year Plan allocated energy and carbon intensity targets to each province. In **India**, several states have submitted action plans (MOEF, 2015). In addition to federal actions and targets to address climate change, **Canada's** provinces and territories have developed their own climate change plans and strategies. Similarly, in **Brazil**, some states are establishing emission reduction targets, with the state of Sao Paulo being the first. In countries with an economy-wide emissions cap, sub-national climate change plans and targets can help raise awareness of climate change and promote climate action (but will not result in additional emission reductions in sectors covered by the cap).

Many cities in the countries studied have also made climate change commitments and implemented various types of mitigation policies (CDP, 2014). For example:

- **Copenhagen** intends to phase out GHG emissions from all sources by 2025.
- **Jakarta** aims to reduce its carbon intensity by 30% by 2030 based on 2005 levels.
- **London** aims to reduce its CO_2 emissions by 60% by 2025 from 1990 levels.
- **Los Angeles** has a target to reduce emissions by 35% from 1990 levels by 2030 in electricity generation and vehicle fuels.
- **Madrid** aims to reduce CO_2 emissions from all sources by 35% compared to 2005 by 2020.
- The **Paris** Climate Plan establishes a target of a 75% reduction in GHG emissions from 2004 levels by 2050 and includes interim targets for 2020, including for energy efficiency and renewable energy.
- **Rio de Janeiro** established a target of 20% reduction by 2020 relative to 2005 levels.
- **Stockholm** has a target to be fossil-fuel free by 2050.
- **Tokyo** has a target to reduce CO_2 emissions by 25% from 2000 levels by 2020.

Carbon pricing and support for fossil fuels

Governments employ a diverse range of policy instruments to meet their climate policy objectives. Many of these instruments cut across multiple policy objectives, economic sectors and levels of government. They include carbon pricing and other mitigation policies in energy and other sectors, such as renewable energy support policies, regulatory standards, innovation and research, development and demonstration (RD&D), and various policies in the agriculture, LULUCF, industry and waste sectors.

Carbon pricing is an essential element of climate change mitigation policy. Placing a price on emissions of CO_2 (and possibly other GHGs) uses market forces to reduce use of GHG-emitting products and services and encourage investment in low-carbon technology. Carbon pricing policies can be cost-effective approaches to reduce emissions, provided markets are well-functioning. Taxes on energy are one form of carbon pricing policy. They include carbon taxes (typically applied per unit of CO_2 emitted) and other taxes on energy.

While energy taxes are widely used, both climate and non-climate policy objectives can influence rates. These objectives include addressing other harmful side effects of energy use such as air pollution. There are many incoherencies in the taxation of energy, with low rates on some of the most harmful fuels, or different tax rates on fuels used for similar purposes in many countries (OECD, 2015e). By contrast, carbon taxes are taxes on energy that have been expressly designed to reflect the CO_2 emissions of different energy sources. Their implementation to date, however, has often proven politically contentious. Where carbon taxes have been established, low tax rates and widespread exemptions (often justified in terms of protecting industrial competitiveness) have limited their impact.

Emissions trading systems also generate carbon prices. Most of these systems, known as cap-and-trade, place a cap on emissions, with tradable permits allocated to installations covered by the cap. Whereas carbon taxes fix the price but not the quantity of emission reductions, emissions trading systems fix the targeted quantity but not the price of emissions allowances. Cap-and-trade provides more certainty regarding environmental outcomes (although setting the cap at the right level is challenging). To date, depressed allowance prices have provided little incentive to undertake the long-term structural changes needed to achieve the transition to a low-carbon economy.

An increasing number of national and sub-national jurisdictions are introducing or raising taxes that put a price on carbon. For example, **Portugal** launched a carbon tax as part of a larger green tax reform that came into force in January 2015 to reduce energy dependence, and encourage sustainable production and consumption. The rate of **Sweden's** carbon tax was increased to EUR 119 per tCO_2 in 2013, although there are significant exemptions. **Mexico** introduced a new tax on fossil fuels based on their carbon content relative to that of natural gas in 2014. **British Columbia** introduced a revenue-neutral tax reform in 2008. The tax, introduced at a rate of CAD 10 per tCO_2e, increased over time to reach CAD 30 per tCO_2e in 2012 (Harrison, 2013). New carbon taxes are also planned in **South Africa and Chile** and an increase in the tax rate is planned in **Alberta** for 2016. Many of these taxes include exemptions and reduced rates for certain fuels or sectors. Considering these explicit taxes jointly with other taxes on energy that implicitly put a price on carbon allows calculation of effective tax rates on carbon (OECD, 2015e).

Governments are pushing ahead with emissions trading systems (ETSs) at supranational, national and sub-national levels. Such systems have now been established in Europe (the **EU ETS**, which applies to the 28 EU Member States and 3 EEA countries), as well as several other national and sub-national jurisdictions. These include **Korea**, **New Zealand** and **Switzerland** at the national level, and **California, Quebec, Tokyo, Saitama,** nine north-eastern **US** states (the Regional Greenhouse Gas Initiative) and seven cities and provinces in **China** at the sub-national level. In January 2015, **Korea** launched an ETS applied to business entities that generate over 125 000 tCO_2e or own facilities generating over 25 000 tCO_2e (three-year annual average rate). **California** and **Quebec** linked their systems in 2014, enabling emissions allowances from either system to be used for compliance in the other.

In **China**, the National Development and Reform Commission (NDRC) is planning to launch a national ETS, building on the seven pilot ETSs. Further systems are under consideration at the national level in **Brazil, Chile, Mexico, Turkey, Japan** and the **Russian Federation**, as well as at the sub-national level in **Canada** and **Brazil**. In most cases, the coverage of ETSs is not economy-wide and limited to large installations in the power and industry sectors.

As surplus of over 2 billion allowances in the **EU ETS** has built up, weakening the strength of the carbon price signal, it is being reformed. As a short-term measure, the European Commission is postponing the auctioning of 900 million allowances until 2019/20 (known as "back-loading"). It has proposed to establish a market stability reserve from 2021 to improve the system's resilience to major shocks: it will adjust the supply of allowances to be auctioned. It has also proposed an increase in the linear reduction factor of the emissions cap from 1.74% to 2.2% per year from 2021 (European Commission, 2015). Further, the low price of allowances in the EU ETS has placed pressure on EU Member States to introduce additional overlapping measures to bolster the carbon price, such as the UK Carbon Floor Price.

However, many countries still have policies in place that directly or indirectly support the production and/or consumption of fossil fuels (OECD, 2015f). By supporting increased use of fossil fuels, these measures undermine the signals that carbon prices are intended to convey. The current environment of low oil and coal prices presents an opportunity to reform measures supporting the consumption of fossil fuels since it alleviates the potential for inflationary impacts. In **Indonesia**, almost 20% of all government spending in 2011 subsidised the consumption of fuel and electricity, disproportionately benefiting richer households. In its revised 2015 budget, however, the Indonesian government entirely phased out gasoline subsidies, leaving only the smaller subsidies for liquefied petroleum gas, diesel fuel and kerosene. In **India**, the central government started reducing consumer subsidies for diesel fuel in late 2012. The savings realised amounted to about INR 200 billion between 2012-14 – roughly equivalent to 10% of the revenues the country derives ever year from all its federal excise duties combined. At the same time, India increased cash transfers for the poor (OECD, 2015f).

Policies to address emissions from power generation and transport

Increased use of low-carbon energy sources such as renewable energy sources, nuclear power and carbon capture and storage (CCS) together with enhanced energy efficiency are needed to decarbonise the energy system. The most common policies used by the countries studied to support renewable energy sources are feed-in tariffs and premiums, renewable portfolio standards (RPSs) and tradable certificates and other fiscal and financial incentives. Feed-in tariffs and premiums reduce risk for eligible generators by guaranteeing a pre-determined price for the electricity generated. Several countries, including **Spain, Italy** and **Germany**, recently reformed the level of feed-in tariffs and renewable energy incentives to improve the financial sustainability of the system. After investigating feed-in tariffs, the **South African** government chose a competitive tender approach that has attracted substantial private sector expertise and investment into grid-connected renewable energy at competitive prices. The largest renewable portfolio standards system is in the **United States**, where 29 states require electric utilities to supply a share (ranging from 2% to 33% by 2020-21) of electricity from renewables.

Regulatory standards are widely used to reduce emissions from vehicles, power plants, buildings and appliances. In the **United States**, emission standards for new and existing power plants are being developed by the US Environmental Protection Agency (EPA) under the Clean Air Act. Emission standards for new and existing coal-fired power plants came into force in **Canada** in 2015. The **European Union** has introduced strict fuel economy standards for new light-duty vehicles. **Japan's** Top Runner programme has been expanded to set efficiency standards for 31 different products, including vehicles, heaters and various electrical appliances, as well as building materials.

Governments are supporting innovation and the research, development and demonstration (RD&D) of new and improved low-carbon technologies. Although investment in energy-related RD&D as a share of GDP remains low, the share of energy-related RD&D

allocated to low-carbon technologies is increasing. Technologies such as energy storage, smart grids, advanced fuels and vehicles, and CCS will be needed to achieve deep and rapid cuts in global GHG emissions. Support for RD&D of CCS remains low, given the significant role this technology is expected to play in reducing global GHG emissions.

Policies to address emissions from sectors other than energy

Agriculture accounts for a significant share of total GHG emissions in some countries studied – over 30% in Ireland and New Zealand, for example. In many countries, the growth rate of agriculture-related GHG emissions has been decoupled from the growth rate of agricultural output. Mitigation policies to date have mainly focused on changes to livestock management, manure management and optimising use of nitrogen fertilisers. Implementing mitigation policies in this sector has proven challenging partly due to the limited availability of low-cost agricultural mitigation technologies in many regions.

Land use, land-use change and forestry range from being a significant source of GHG emissions in some countries to a significant sink in others. For example, CO_2 removals from Latvia's large forested areas are greater than its GHG emissions from all other sectors, making the country a net GHG sink. In other countries, such as Brazil and Indonesia, deforestation can be a major source of GHG emissions, although deforestation rates in Brazil have been reduced in recent years from high historical levels.

GHG emissions from industrial activity stem from industrial processes, as well as from energy use. Some industrial emissions are a necessary by-product of production (e.g. process CO_2 from clinker production), whereas others can be more easily reduced (e.g. N_2O from nitric acid production). Emissions from industrial processes have decreased since 1990 in OECD member countries, but still account for 8% of national emissions on average. Policies being used to decrease GHG emissions from industry include economic instruments, regulations and information programmes. Industries in some countries are also taking voluntary action to mitigate GHG emissions.

GHG emissions from waste, which represent national emissions of between 0.4% in Luxembourg and 12% in Portugal, are being addressed mainly through fiscal incentives (mostly landfill taxes) and regulations. Mitigation options in the waste sector include waste prevention measures, promotion of waste recycling and recovery, and capture and combustion of landfill gas. Important mitigation potential lies in measures dealing with the entire life cycle – from production to consumption to end-of-life of materials (OECD, 2012).

References

CBI China (2012), "Translation of the 12th Five-Year Plan of China", *www.cbichina.org.cn/cbichina/upload/fckeditor/Full%20Translation%20of%20the%2012th%20Five-Year%20Plan.pdf.*

CDP (2014), *Cities Emissions Reduction Targets in 2014*, Carbon Disclosure Project, *https://data.cdp.net/Cities/Cities-Emissions-Reduction-Targets-2014/vk38-i4tf* (accessed 20 July 2015).

Conseil Constitutionnel (2013), "Communiqué de presse : décision du Conseil", *www.conseil-constitutionnel.fr/conseil-constitutionnel/francais/les-decisions/acces-par-date/decisions-depuis-1959/2013/2013-346-qpc/communique-de-presse.138284.html.*

European Commission (2015), Structural Reform of the European Carbon Market website, *http://ec.europa.eu/clima/policies/ets/reform/index_en.htm* (accessed 23 June 2015).

Finland Ministry of the Environment (2015), "Parliament approves the Climate Change Act", webpage, *www.ym.fi/en-US/Latest_news/Press_release_service/Parliament_approves_the_Climate_Change_A(32988)* (accessed 20 July 2015).

France Ministry of Foreign Affairs and International Development (2014), "Climate disruption/people's climate march", 20 September, *www.ambafrance-uk.org/Foreign-Minister-issues-stark* (accessed 07 July 2015).

Government of Canada (2014), *Canada's Sixth National Report on Climate Change to the United Nations Framework Convention on Climate Change*, *http://unfccc.int/national_reports/annex_i_natcom_/items/1095.php*.

Government of France (2015), The Agenda of Solutions or the Lima-Paris Action Plan website, *www.cop21.gouv.fr/en/media-facilities/press-room/decryptage-lagenda-des-solutions-ou-le-plan-daction-lima-paris* (accessed 23 June 2015).

Government of India (2008), *National Action Plan on Climate Change*, *www.moef.nic.in/downloads/home/Pg01-52.pdf*.

Harrison, K. (2013), "The political economy of British Columbia's carbon tax", *OECD Environment Working Papers*, No. 63, OECD Publishing, Paris, *http://dx.doi.org/10.1787/5k3z04gkkhkg-en*.

IEA (2015a), *World Energy Outlook 2015 Special Report on Energy and Climate*, IEA/OECD Publishing, Paris.

IEA (2015b) "World energy balances", *IEA World Energy Statistics and Balances* (database), *http://dx.doi.org/10.1787/data-00512-en* (accessed 21 July 2015).

IEA (2015c), "Indicators for CO_2 emissions", *IEA CO_2 Emissions from Fuel Combustion Statistics* (database), *http://dx.doi.org/10.1787/data-00433-en* (accessed 21 July 2015).

IEA (2014a), *World Energy Investment Outlook 2014*, IEA/OECD Publishing, Paris, *https://www.iea.org/publications/freepublications/publication/WEIO2014.pdf*.

IEA (2014b) *World Energy Outlook 2014 – Electricity Access Database*, *www.worldenergyoutlook.org/media/weowebsite/WEO2014Electricitydatabase(1).xlsx* (accessed 25 April 2015).

IEA (2014c), *Capturing the Multiple Benefits of Energy Efficiency: A Guide to Quantifying the Value Added*, IEA/OECD Publishing, Paris, *http://dx.doi.org/10.1787/9789264220720-en*.

IPCC (2014), "Climate change 2014: Mitigation of climate change", in *Fifth Assessment Report of the Intergovernmental Panel on Climate Change*, Cambridge University Press, Cambridge, United Kingdom and New York, US.

IRENA (2015), "Renewable Energy Targets Quadrupled Globally Since 2005, New IRENA Report Finds", Press Release, *www.irena.org/News/Description.aspx?NType=A&mnu=cat&PriMenuID=16&CatID=84&News_ID=413*.

JAIF (2015) "Nuclear power after Fukushima", Japan Atomic Industrial Forum, *www.jaif.or.jp/en/slides/nuclear-power-after-fukushima-2/*.

MOEF (2015), *State Action Plan on Climate Change, Ministry of Environment, Forest and Climate Change*, Government of India, *www.moef.nic.in/ccd-sapcc* (accessed 20 April 2015).

OECD (2012), *Greenhouse Gas Emissions and the Potential for Mitigation from Materials Management within OECD Countries*, OECD, Paris, *www.oecd.org/env/waste/50035102.pdf*.

OECD (2015a), *OECD Embodied Carbon Dioxide Emissions*, edition 2015, *www.oecd.org/sti/inputoutput/co2* (accessed 20 July 2015).

OECD (2015b), *OECD Economic Surveys: Estonia 2015*, OECD Publishing, Paris, *http://dx.doi.org/10.1787/eco_surveys-est-2015-en*.

OECD (2015c), *OECD Economic Surveys: Japan 2015*, OECD Publishing, Paris, *http://dx.doi.org/10.1787/eco_surveys-jpn-2015-en*.

OECD (2015d), *Towards Green Growth? Tracking Progress*, OECD Green Growth Studies, OECD Publishing, Paris, *http://dx.doi.org/10.1787/9789264234437-en*.

OECD (2015e), *Taxing Energy Use 2015: OECD and Selected Partner Economies*, OECD Publishing, Paris, *http://dx.doi.org/10.1787/9789264232334-en*.

OECD (2015f), *OECD Companion to the Inventory of Support Measures for Fossil Fuels 2015*, OECD Publishing, Paris, *http://dx.doi.org/10.1787/9789264239616-en*.

OECD (2015g) "Greenhouse gas emissions by source", *OECD Environment Statistics* (database) based on *National Inventory Submissions 2014* to the *United Nations Framework Convention on Climate Change* (UNFCCC, CRF tables), and replies to the OECD State of the Environment Questionnaire, *http://dx.doi.org/10.1787/data-00594-en* (accessed 20 July 2015).

OECD (2014a), *OECD Economic Surveys: United States 2014*, OECD Publishing, Paris, *http://dx.doi.org/10.1787/eco_surveys-usa-2014-en*.

OECD (2014b), *OECD Economic Surveys: India 2014*, OECD Publishing, Paris, *http://dx.doi.org/10.1787/eco_surveys-ind-2014-en*.

OECD (2014c), *The Cost of Air Pollution: Health Impacts of Road Transport*, OECD Publishing, Paris, *http://dx.doi.org/10.1787/9789264210448-3-en*.

OECD, IEA, ITF, NEA (2015), *Aligning Policies for a Low-carbon Economy*, OECD Publishing, Paris, *http://dx.doi.org/10.1787/9789264233294-en*.

REN21 (2015), *Renewables 2015 Global Status Report*, Renewable Energy Network for the 21st Century, Paris.

UNEP (2015), *Climate Commitments of Subnational Actors and Business: A Quantitative Assessment of their Emission Reduction Impact*, United Nations Environment Programme, Nairobi.

UNFCCC (2015), GHG inventory data detailed by Party, *http://unfccc.int/di/DetailedByParty.do* (accessed 7 July 2015).

UNFCCC (2010), "Report of the Conference of the Parties on its fifteenth session", Copenhagen, 7-19 December 2009, United Nations Framework Convention on Climate Change, *http://unfccc.int/resource/docs/2009/cop15/eng/11a01.pdf*.

UNFCCC (1992), *United Nations Framework Convention on Climate Change*, *http://unfccc.int/files/essential_background/background_publications_htmlpdf/application/pdf/conveng.pdf*.

Chapter 2

Targets and goals for climate change mitigation

This chapter presents targets and goals that countries have made in the context of the United Nations Framework Convention on Climate Change (UNFCCC) to limit or reduce their greenhouse gas (GHG) emissions. It presents Kyoto Protocol commitments, mitigation pledges for 2020 and intended nationally determined contributions (INDCs). It provides a simple analysis of these targets and goals. It also presents climate-related national targets such as increasing the use of renewable energy, reducing energy consumption and increasing forest coverage or volume.

The statistical data for Israel are supplied by and under the responsibility of the relevant Israeli authorities. The use of such data by the OECD is without prejudice to the status of the Golan Heights, East Jerusalem and Israeli settlements in the West Bank under the terms of international law.

Targets and goals for 2020 and beyond

As part of multilateral efforts to address climate change, countries have submitted various types of targets and goals to reduce or limit their GHG emissions. These targets and goals have different timescales and legal statuses. They include emission reduction commitments that are internationally binding for some Annex I Parties under the Kyoto Protocol; mitigation pledges by developed and developing countries for 2020 under the UNFCCC; and INDCs from all countries for after 2020. These are summarised in turn below.

Kyoto Protocol commitments

The Kyoto Protocol, signed in 1997, set out emission reduction commitments for Annex I Parties that had ratified the UNFCCC at the time of negotiating the Kyoto Protocol (i.e. all except Turkey). The first commitment period ran from 2008-12 and the second commitment period runs from 2013-20. Commitments under the Kyoto Protocol are expressed in the form of emission budgets over a period. Countries can meet their commitments by (i) reducing GHG emissions domestically; (ii) enhancing domestic CO_2 removals by forests; and/or (iii) purchasing emission offsets from designated international carbon markets.

Table 2.1 summarises commitments taken under the first and second commitment periods of the Kyoto Protocol. Of the countries studied in this report, 30 took on commitments for the first period (including 23 EU Member States, which chose to fulfil their commitment jointly) and 27 took on commitments for the second period. Canada took on a commitment for the first period, but subsequently withdrew from the Kyoto Protocol in 2011. Japan, New Zealand and the Russian Federation took on commitments for the first period, but not the second one. The United States is not a Party to the Kyoto Protocol. Turkey had not ratified the convention at the time of negotiating the Kyoto Protocol so did not have a target.

Mitigation pledges for 2020

In 2009 and 2010, Parties to the UNFCCC were invited to put forward mitigation pledges for 2020. These pledges were known as "quantified economy-wide emissions targets" for developed countries and "nationally appropriate mitigation actions" (NAMAs) for developing countries (UNFCCC, 2014, 2013). Developed countries expressed pledges as absolute emission reductions relative to a base year, while developing countries expressed pledges using a variety of different metrics. These included reductions in emission intensity from a base year, reductions in total GHG emissions relative to a business-as-usual (BAU) baseline and achievement of carbon neutrality. Further, some developing countries submitted sector-specific targets for increasing the use of renewable energy sources, increasing use of biofuels and/or enhancing forest cover and stock volume (Table 2.2). Of the countries

Table 2.1. **Kyoto Protocol commitments**

Party	Quantified emission limitation or reduction commitment for first commitment period (2008-12, from 1990 levels unless specified otherwise)	Quantified emission limitation or reduction commitment for second commitment period (2013-20)
Australia	+8%	–0.5% from 1990 levels[a]
Austria	–13%[b]	–16% from 2005 levels[e]
Belgium	–7.5%[b c]	–15% from 2005 levels[e]
Canada	–6%[d]	N/A
Czech Republic	–8%[c]	+9% from 2005 levels[c]
Denmark	–21%[b c]	–20% from 2005 levels[e]
Estonia	–8%[c]	+11% from 2005 levels[e]
EU	–8%	–20% from 1990 levels
Finland	0%[b c]	–16% from 2005 levels[e]
France	0%[b]	–14% from 2005 levels[e]
Germany	–21%[b c]	–14% from 2005 levels[e]
Greece	+25%[b c]	–4% from 2005 levels[e]
Hungary	–6% from 1985–87 levels[c]	+10% from 2005 levels[e]
Iceland	+10%[b]	–20% from 1990 levels[e]
Ireland	+13%[b c]	–20% from 2005 levels[e]
Italy	–6.5%[b]	–13% from 2005 levels[e]
Japan	–6%[c]	N/A
Latvia	–8%[c]	+17% from 2005 levels[e]
Lithuania	–8%[c]	+15% from 2005 levels[e]
Luxembourg	–28%[b c]	–20% from 2005 levels[e]
Netherlands	–6%[b c]	–16% from 2005 levels[e]
New Zealand	0%	N/A
Norway	+1%	–16% from 1990 levels
Poland	–6% from 1988 levels[c]	+14% from 2005 levels[e]
Portugal	+27%[b c]	+1% from 2005 levels[e]
Russian Federation	0%[c]	N/A
Slovak Republic	–8%	+13% from 2005 levels[e]
Slovenia	–8% from 1986 levels[c]	+4% from 2005 levels[e]
Spain	+15%[b c]	–10% from 2005 levels[e]
Sweden	+4%[b c]	–17% from 2005 levels[e]
Switzerland	–8%	–15.8% from 1990 levels
United Kingdom	–12.5%[b c]	–16% from 2005 levels[e]

[a] Australia unconditionally pledged to reduced its emissions by 5% by 2020 from 2000 levels. This pledge was translated into a quantified emission limitation or reduction commitment of –0.5% from 1990 levels over 2013-20.
[b] As defined by the EU burden-sharing agreement (Council Decision 2002/358/EC).
[c] With a base year of 1995 for F-gases.
[d] Canada withdrew from the Kyoto Protocol in 2011.
[e] For emissions from non-ETS sectors only, as defined by the EU Effort Sharing Decision (Decision 406/2009/EC).

Source: UNFCCC (2015c), Kyoto Protocol website, *United Nations Framework Convention on Climate Change*, *http://unfccc.int/kyoto_protocol/items/2830.php*.

studied in this report, all countries except Turkey have made mitigation pledges for 2020 under the UNFCCC.

Some mitigation pledges for 2020 have conditions attached. For example, some are conditional on significant action by other major emitters, on the agreement of an ambitious global climate change deal, on the use of carbon market mechanisms or on financial support (for some developing countries). Further, several countries submitted a range of pledges rather than a single emission reduction pledge, often in combination with the conditions under which the country would undertake its most ambitious pledge.

Table 2.2. **Mitigation pledges for 2020**

Party	Mitigation pledges for 2020
Australia*	Reduce GHG emissions by -5% (unconditional), -5% to -15% (conditional) or -25% (conditional) from 2000 levels
Brazil	Reduce GHG emissions in the range of 36.1% and 38.9% below its projected emissions
Canada	Reduce GHG emissions by -17% from 2005 levels
Chile	Reduce GHG emissions by -20% below BAU
China	Reduce CO_2 per unit of GDP by 40-45% from 2005 levels Increase share of non-fossil fuels in primary energy consumption to 15% Increase forest coverage by 40 million ha and forest stock volume by 1.3 billion m^3 from 2005 levels
Colombia	Increase renewable energy installed capacity to 77% of total installed capacity Reduce deforestation in the Colombian Amazon rainforest to zero Increase consumption of biofuels to 20% of total fuel consumption
Costa Rica	Become carbon neutral, starting year 2021
EU*	Reduce GHG emissions by -20% (unconditional) or -30% (conditional) from 1990 levels
Iceland	Reduce GHG emissions by -15% (unconditional) or -30% (joint effort with the EU) from 1990 levels
India	Reduce GHG emissions per unit GDP by 20-25% from 2005 levels (excluding agriculture)
Indonesia	Reduce GHG emissions by 26% below BAU unilaterally or 41% below BAU with support
Israel	Reduce GHG emissions by -20% below BAU Increase share of renewable energy in electricity generation to 10% Decrease electricity consumption by 10%
Japan	Reduce GHG emissions by -3.8% from 2005 levels
Korea	Reduce GHG emissions by -30% below BAU
Mexico	Reduce GHG emissions by up to 30% compared with the BAU scenario
New Zealand	Reduce GHG emissions by -5% (unconditional) or -10% to -20% (conditional) from 1990 levels
Norway*	Reduce GHG emissions by -30% (unconditional) or -40% (conditional) from 1990 levels
Russian Federation	Reduce GHG emissions by -15% to -25% from 1990 levels
South Africa	Reduce GHG emissions by 34% below BAU by 2020 and 42% below BAU by 2025
Switzerland*	Reduce GHG emissions by -20% or -30% from 1990 levels

* Australia, the European Union, Iceland, Norway and Switzerland translated their mitigation pledges for 2020 into emission reduction commitments under the second commitment period of the Kyoto Protocol.

Source: UNFCCC (2011a), *Compilation of economy-wide emission reduction targets to be implemented by Parties included in Annex I to the Convention, http://unfccc.int/resource/docs/2011/sb/eng/inf01r01.pdf;* UNFCCC (2011b), *Compilation of information on nationally appropriate mitigation actions to be implemented by Parties not included in Annex I to the Convention, http://unfccc.int/resource/docs/2011/awglca14/eng/inf01.pdf.*

Intended nationally determined contributions for after 2020

At COP 19 in Warsaw, Parties to the UNFCCC agreed to put forward INDCs for the post-2020 period (summarised in Table 2.3). There is no internationally-agreed form in which INDCs are to be expressed, although each is to represent "a progression beyond the current undertaking of that Party" (UNFCCC, 2015a). The accounting rules that Parties intend to use vary (Box 2.1). Further, countries are encouraged to provide information to clarify the nature of their INDC. The level of information provided by countries will affect the ease with which the INDCs can be translated into expected future GHG emissions levels. This information is needed to assess future global emissions pathways and the corresponding probabilities of different global average temperature increases.

National climate change targets, plans and general approaches

In addition to their international mitigation targets and goals to limit or reduce emissions, many countries have set domestic targets in areas such as renewable energy and energy efficiency as part of national planning processes. Targets enable countries to structure their short-term actions within a long-term goal. These processes can provide an opportunity to diagnose misalignments between climate goals and overall policy and regulatory frameworks and improve co-ordination among various stakeholders, including government ministries and the private sector (OECD/IEA/NEA/ITF, 2015).

Table 2.3. **Intended nationally determined contributions**

Party	Metric	Headline number	Base year	End year
Australia	Total GHG emissions	26-28% reduction	2005	2030
Brazil*	Total GHG emissions	37% reduction	2005	2025
Canada	Total GHG emissions	30% reduction	2005	2030
Chile*	CO_2 intensity	30% reduction	2007	2030
China	CO_2 emissions	Peak CO_2 emissions by 2030, "making best efforts to peak early"	-	2030
	CO_2 intensity	60-65% reduction	2005	2030
	Share of non-fossil fuels in primary energy consumption	Increase to around 20%	-	2030
	Forest stock volume	Increase by around 4.5 billion m^3	2005	2030
Colombia*	Total GHG emissions	20% reduction unconditionally, or up to 30% with support	BAU	2030
Costa Rica*	Total net GHG emissions	Limit to a maximum of 9.374 $MtCO_2e$	-	2030
EU-28	Total GHG emissions	At least 40% reduction (to be achieved domestically)	1990	2030
Iceland	Total GHG emissions	To be determined (aims to be delivered collectively with the EU)	1990	2030
India*	Emissions intensity	33-35% reduction	2005	2030
Indonesia*	Total GHG emissions	29% reduction unconditionally, or up to 41% with support	BAU	2030
Israel*	GHG emissions per capita	Reduce to 7.7 tCO_2e per capita	-	2030
Japan	Total GHG emissions	26% reduction	2013	2030
Korea	Total GHG emissions	37% reduction	BAU	2030
Mexico	Total GHG emissions	25% reduction unconditionally, or up to 40% with support	BAU	2030
New Zealand	Total GHG emissions	30% reduction	2005	2030
Norway	Total GHG emissions	At least 40% reduction (to be delivered collectively with the EU)	1990	2030
Russian Federation	Total GHG emissions	25-30% reduction (subject to accounting rules for forests)	1990	2030
South Africa*	Total GHG emissions	To be in the range between 398 and 614 $MtCO_2e$ by 2025 and 2030 (peak, plateau and decline trajectory range)	-	2025/2030
Switzerland	Total GHG emissions	50% reduction	1990	2030
Turkey*	Total GHG emissions	Up to 21% reduction	BAU	2030
United States	Total GHG emissions	26-28% reduction (with "best efforts" to reduce emissions by 28%)	2005	2025

*INDCs submitted after August 2015.

Note: Other countries that are not covered in this report have also submitted INDCs to the UNFCCC.

Source: UNFCCC (2015b), *"INDCs as communicated by Parties", web portal, United Nations Framework Convention on Climate Change, www.4.unfccc.int/submissions/indc/Submission%20Pages/submissions.aspx* (accessed 20 July 2015).

Some countries are establishing clear legal frameworks to reach their targets, including national carbon budgets or targets for cumulative emissions. For instance, through its 2008 Climate Change Act, the **United Kingdom** institutionalised carbon budgets that restrict the total amount of GHG emissions that the country can emit over a five-year period. This contributes to the long-term goal of reaching at least 80% reduction by 2050. The UK carbon budgets, which unfold in four periods until 2027, are the first legally binding carbon budgets (Government of the United Kingdom, 2014).

In **Denmark**, a Climate Change Act was passed by the Danish Parliament in 2014 that establishes a strategic framework for Denmark's climate policy and a system for setting emission reduction targets with a ten-year perspective once every five years. The **Norwegian** Parliament has requested the government to develop a proposal for climate legislation that

Box 2.1. Accounting for mitigation targets and goals

In many cases, progress in implementation of mitigation targets and goals cannot be assessed simply by looking at the GHG inventory of the Party concerned. This is because Parties can choose to use GHG units from market mechanisms (corresponding to mitigation occurring outside the scope of the GHG inventory) to meet part of their targets and goals. Further, the approach used to account for GHG emissions and removals from land use, land-use change and forestry (LULUCF) may differ to that used for compiling the GHG inventory. The rationale for introducing such flexibility into accounting is to enable Parties to enhance the cost effectiveness of meeting their mitigation targets and goals and, in the case of LULUCF, ensure that only the results of anthropogenic actions are counted towards the achievement of mitigation targets and goals.

A clear set of accounting rules was developed for the first and second commitment periods of the Kyoto Protocol. To meet their Kyoto Protocol commitments, Annex I Kyoto Protocol Parties are to ensure GHG emissions do not exceed their holdings of tradable GHG units. These units include emission offsets from the Clean Development Mechanism. Participating Parties have not agreed on a common set of accounting rules for their targets and goals. Therefore, accounting approaches to assess progress in implementation vary considerably. In particular, it remains unclear which tradable GHG units can be used by Parties to meet part of their 2020 targets and goals.

Countries have taken different approaches to the use of market mechanisms and accounting for LULUCF in the context of their INDCs. For example, some Parties (e.g. the European Union, the Russian Federation and the United States) intend to achieve their INDCs through domestic reductions only, while other Parties (e.g. Korea, Switzerland) intend to use emission credits to meet part of their contributions. The European Union has announced it will establish its approach to account for LULUCF "as soon as technical conditions allow and in any case before 2020" (EU, 2015). Norway's approach for LULUCF accounting will be developed in consultation with the European Union (in the case of collective delivery with the EU); Norway has stated its final choice of land sector accounting shall not affect the ambition level of its INDC. The United States will use its GHG inventory reporting as the basis for LULUCF accounting. The Russian Federation's INDC to reduce GHG emissions by 25-30% compared to 1990 levels is conditional upon the use of accounting rules that maximise recognition of removals by the LULUCF sector. New Zealand's INDC will remain provisional pending confirmation of the approaches to be taken in accounting for the land sector, and confirmation of access to carbon markets.

Japan intends to use credits from a Joint Crediting Mechanism (JCM) to meet part of its emissions reduction targets. The JCM facilitates the diffusion of leading low-carbon technologies, products, systems, services and infrastructure, as well as implementation of mitigation actions. It also aims to contribute to achieving sustainable development in developing countries. As of June 2015, Japan has consulted for the JCM with developing countries since 2011 and signed bilateral JCM agreements with 14 countries (Japan, 2015).

establishes national targets for emission reductions for 2030 and 2050. **Finland's** Parliament approved a Climate Change Act in 2015 that establishes an emission reduction target of 80% from 1990 levels by 2050, with long-term mitigation action plans to be drawn up every ten years (Finland Ministry of the Environment, 2015). **France** adopted an Energy Transition Law in 2015 that sets targets to reduce GHG emissions by 40% by 2030 from 1990 levels, divide emissions by four by 2050, halve energy consumption by 2050, cap the total output from nuclear power plants and increase the share of renewables to 32% of total final consumption by 2030.

Canada's general approach to climate change mitigation policies is to regulate emissions on a sector-by-sector basis. To this end, the Canadian government has announced its intention to develop new regulations to address methane emissions in the oil and gas sector (aligned with action in the United States), GHG emissions from natural gas-fired electricity generation and production of chemicals and fertilisers, hydrofluorocarbons (HFCs) and post-2018 heavy-duty vehicles. **Australia** has taken a direct approach to meeting its 2020 emissions reduction target by establishing an Emissions Reduction Fund that purchases abatement as it is delivered by projects across the economy. **Germany** has developed a comprehensive climate change mitigation programme in which all sectors must contribute towards the EU target of a 40% reduction below 1990 levels by 2030. Approximately three-quarters of reductions are projected to come from the National Action Plan on Energy Efficiency (including buildings, but excluding transport) and the electricity sector.

Most countries have overarching climate change plans or strategies that complement international targets. Examples follow below.

- **China** has formalised efforts to mitigate climate change in recent five-year plans. For example, the 12th Five-Year Plan for National Economic and Social Development (2011-15), expressed the need to pursue low-carbon development. It established a binding target of a 17% decrease in CO_2 emissions per unit of GDP from 2010 levels by the end of 2015. The plan includes several policy measures, such as gradually establishing carbon emissions trading systems (ETSs). China's State Council has also announced that annual coal consumption will be capped at below 4.2 billion tonnes until 2020 (Xinhuanet, 2014). China's coal consumption decreased by 2.9% between 2013-14 (National Bureau of Statistics of China, 2015).
- The **European Union's** international target has been translated into an EU-wide policy framework. Based on the experience gained to reach its 20/20/20 targets, the European Commission developed a proposal for the EU 2030 Policy Framework for Climate and Energy, which was adopted in October 2014. It includes a set of new targets for 2030 on energy efficiency, renewable energy and GHG emission reductions. The EU set a binding target of reducing GHG emissions by 40% below 1990 levels by 2030. In contrast to the 2020 target, the 2030 target will be met through domestic measures only, with no use of international offsets. The EU also established an EU-wide binding target to boost the share of renewable energy to at least 27% of the EU's energy consumption by 2030. This EU-wide target, which will be reviewed in 2020, will not be translated into legally binding renewables targets for individual Member States beyond 2020. Moreover, energy efficiency targets will not be translated into nationally binding targets.
- As part of the National Action Plan on Climate Change (2008), **India** sought to promote development objectives, while addressing climate change. The action plan outlines eight national missions for 2017 regarding energy efficiency, solar technology, sustainable habitats, water, Himalayan ecosystems, "green India", agriculture and strategic knowledge (Government of India, 2012). In 2010, India launched the Jawaharlal Nehru National Solar Mission (JNNSM) with the aim of establishing India as a global hub for solar manufacturing and increasing the share of solar energy in the total energy supply. The JNNSM sets a policy framework for deploying 20 gigawatts (GW) of grid-connected solar photovoltaic (PV) and 2 GW of off-grid PV by 2022, with the aim of increasing energy access in remote areas. In November 2014, it was announced the target would be increased to 100 GW, with 60 GW of large and medium grid-connected solar PV and 40 GW rooftop systems.
- In 2009, the **United States** committed to reduce emissions by 17% below 2005 levels by 2020. Subsequently, President Barack Obama set out a Climate Action Plan in 2013 that encompasses efforts in mitigation and adaptation in the United States, as well as internationally (Executive Office of the President, 2013). The Obama administration outlined actions to support clean energy and low-carbon transport, as well as to reduce non-CO_2 emissions. More than half of US states have renewable energy and energy efficiency targets, which contribute to decarbonising the energy sector. The administration supports the development of renewable energy with the goal of doubling renewable electricity generation by 2020. The action plan also allocates approximately

USD 7.9 billion of the 2014 fiscal year budget to research and development for lower carbon technologies such as nuclear power, as well as efficient use of coal.

Over 160 countries have established renewable energy targets, which can apply to electricity generation, total primary energy supply or energy consumption (IRENA, 2015). In most countries, the highest contribution of renewable energy in total primary energy supply comes from hydropower, biofuels and waste (Figure 2.1). This includes

Figure 2.1. **Renewable energy in electricity generation**

Percentage of electricity generation and share of renewable energy source of electricity, 2013*

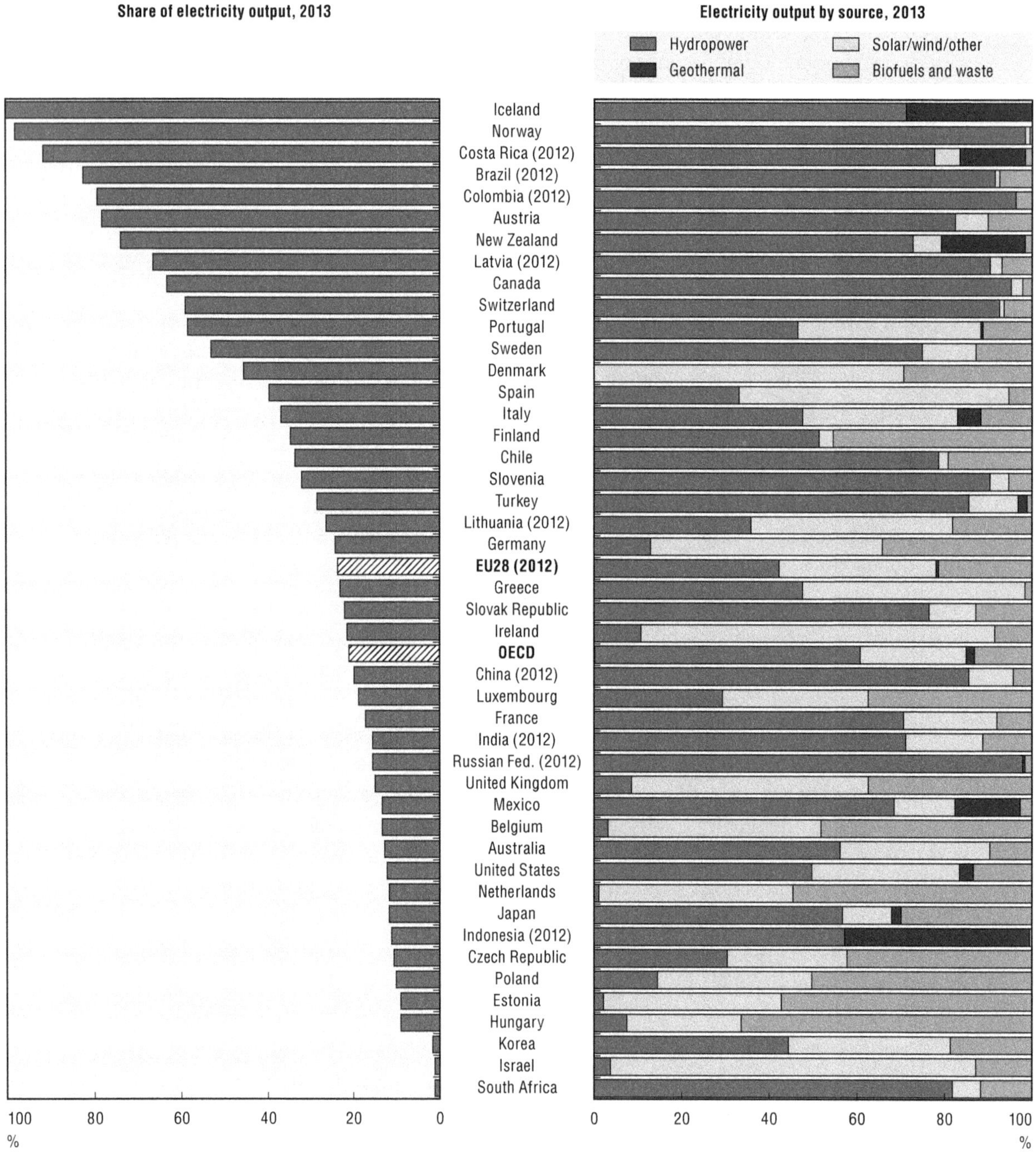

Source: IEA (2015a), IEA World Energy Statistics and Balances (database), *http://dx.doi.org/10.1787/data-00512-en.*

StatLink *http://dx.doi.org/10.1787/888933272456*

primary solid biofuels – any plant matter used directly as fuel, a sub-category that covers various woody materials generated by industrial process or provided directly by forestry and agriculture. Primary solid biofuels account for the main share of the category "biofuels and waste" for Brazil (85%), China (94%), India (99%) and Indonesia (99%). In some of these countries, however, "biofuels and waste" can take the form of traditional solid biofuels that may not be sustainable and can cause indoor air pollution. In 2012, renewable energy accounted for more than a third of total primary energy supply in Iceland (90%), Costa Rica (52%), Brazil (41%), Latvia (37%) and Indonesia (33%). Countries with the highest share of renewables in electricity output were Iceland (100%), Norway (98%), Costa Rica (92%), Brazil (82%) and Colombia (80%). Relative to annual GDP, Costa Rica was among the top countries for investment in new renewable power and fuels.

In addition to renewable energy targets (Table 2.4), some countries, states or municipalities have set targets for certain sectors or activities. The public sector in British Columbia, for instance, has achieved carbon neutrality for five consecutive years (Government of British Columbia, 2014). This commitment covers the entire provincial public sector, including schools, post-secondary institutions, government offices, Crown corporations and hospitals. The US Environmental Protection Agency (EPA) introduced a target in January 2015 to cut

Table 2.4. **Renewable energy targets**

EU Member States

EU member states	Share of renewables in gross final energy consumption, 2013 (%)	Target share of renewables in gross final energy consumption, 2020 (%)
Austria	32.6	34
Belgium	7.9	13
Czech Republic	12.4	13
Denmark	27.2	30
Estonia	25.6	25
Finland	36.8	38
France	14.2	23
Germany	12.4	18
Greece	15.0	18
Hungary	9.8	13
Ireland	7.8	16
Italy	16.7	17
Latvia*	37.1	40
Lithuania*	23.0	23
Luxembourg	3.6	11
Netherlands	4.5	14
Poland	11.3	15
Portugal	25.7	31
Slovak Republic	9.8	14
Slovenia	21.5	25
Spain	15.4	20
Sweden	52.1	49
United Kingdom	5.1	15
EU	15	20

Other countries

Countries	Target share of renewables in electricity generation (unless otherwise specified)
Australia	23.5% by 2020
Brazil*	28-33% by 2030 (excluding hydro power)
Canada	No national target, but nine provincial targets
Chile	20% by 2025
China*	15% non-fossil fuels of primary energy consumption by 2020; increase the non-fossil fuel share to around 20% by 2030
Colombia*	6.5% by 2020 on-grid systems
Iceland	72% by 2020
India*	10% of new capacity additions by 2012
Indonesia*	23% in TPES by 2025
Israel	10% by 2020
Japan	Over 13.5% by 2020
	Over 20% by 2030
Korea	11% in TPES by 2030 (new and renewable energy)
Mexico	5% by 2018
	35% by 2024 (clean sources)
New Zealand	90% by 2025
Norway	67.5%
Russia*	2.5%
Switzerland	Annual production must increase by at least 5 400 GWh compared to 2000 by 2030
Turkey	30% by 2023
United States	The RPS requirements vary by state (from 3% by 2021 to 33% by 2020)

*Partner economies.

Source: European Commission (2009); METI (2014), "Strategic energy plan", Ministry of Economy, Trade and Industry, Japan, April 2014, *www.enecho.meti.go.jp/en/category/others/basic_plan/pdf/4th_strategic_energy_plan.pdf*; REN21 (2015) "Renewable energy interactive map".

methane emissions from oil and gas production by 40-45% by 2025 below 2012 levels; it also outlined actions to achieve this target, building on prior activities by the administration (White House, 2015). Action is also being taken to address emissions from the buildings sector; the United Kingdom, for instance, aims to equip all homes and small businesses with smart meters by 2020.

Analysis of emission reduction rates needed to meet mitigation targets and goals

This sub-section analyses the international targets and goals outlined above. It considers the Kyoto Protocol commitments and mitigation pledges for 2020, as well as the INDCs that have been announced for 2030. The aim is to compare where countries need to be with where they have recently been, based on the latest statistics available. An overview of the methodology and individual country results are provided in Annex A. The *World Energy Outlook Special Report on Energy and Climate* (IEA, 2015b) provides an in-depth analysis of INDCs with a focus on the energy sector.

First, the analysis calculates the average annual rates of change of total GHG emissions (or CO_2 intensity in the case of China and India) needed to meet the targets and goals. In cases where countries have submitted a range of pledges, a range of annual rates is provided. These rates are compared to historical rates in 1990-2000, 2000-05 and 2005-12, as well as maximum and minimum values observed since 1990 (Table 2.5). The results show that some countries will need unprecedented average annual rates of change in total GHG emissions to meet their targets and goals. In other countries, the rates of change of emissions needed to meet targets and goals do not exceed historically observed values. In these cases, however, meeting the targets and goals may well still prove challenging, particularly if strong action was taken in the past, if the economy is rapidly growing or if external factors such as economic recession rather than climate policies caused past decreases in GHG emissions.

Second, the analysis considers total GHG emissions as the product of two drivers: (i) economic output (GDP); and (ii) GHG emissions intensity per unit of economic output (kgCO_2e per unit of GDP). Historical annual rates of change in these drivers for all countries studied since 1990 (subject to data limitations) are shown in Figure 2.2: since 1990, many countries have been decreasing emission intensity while increasing GDP. Nevertheless, the reduction rates of average annual emission intensity will need to be accelerated to achieve deep cuts in total GHG emissions. Reduction rates, in turn, depend on the GDP growth rate, with higher growth scenarios implying stronger annual reductions in emission intensity.

There are some important caveats to this simple analysis. First, full-time series of GHG statistics (1990-2012) are not available for all countries. Second, many of the analyses exclude emissions and removals from the LULUCF sector; it remains unclear how mitigation pledges and INDCs will account for this sector. Third, the analysis excludes use of offsets from market mechanisms. In reality, several countries intend to use carbon credits from international market mechanisms to meet part of their mitigation targets and goals. This would decrease the rates of domestic emission reduction required to meet any given mitigation target or goal.

Table 2.5. **Analysis of country mitigation targets and goals**

Country	LULUCF	Historical annual rates of change (%)				Future annual rates of change needed to meet mitigation targets and goals (%)	
		1990-2000	2000-05	2005-12	Max/Min	2020 pledges	Post-2020 INDCs
Annual rates of change of total GHG emissions, emission reduction targets relative to a base year ($MtCO_2e$)							
Australia	Excluded	+1.5	+1.4	+0.5	+3.0/-0.7	-	-
	Included	-0.1	+1.8	-1.2	+4.8/-5.7	-0.4 to -3.7	-1.6 to -1.9
Canada	Excluded	+2.0	+0.4	-0.7	+3.1/-4.1	-1.7	-1.7
	Included	+2.6	+3.3	-0.9	+21.4/-14.9	-1.5	-1.7
EU-28	Excluded	-0.9	+0.2	-1.8	+1.5/-4.8	-0.1 to -1.8	-2.8
	Included	-1.1	+0.2	-2.0	+1.2/-5.7	+0.2 to -1.5	-2.8
Iceland	Excluded	+1.0	-0.2	+2.1	+9.4/-3.8	-4.8 to -7.1	See EU-28
	Included	+0.4	-0.6	+1.2	+7.3/-3.9	+3.1 to -5.5	See EU-28
Japan	Excluded	+0.8	+0.1	-0.1	+4.1/-6.0	-0.4	-2.2
	Included	+0.7	+0.1	+0.1	+4.2/-6.0	-0.6	-1.5
New Zealand	Excluded	+1.6	+2.0	-0.4	+3.6/-2.1	-3.4 to -5.5	-0.5
	Included	+5.1	+4.6	+0.4	+9.6/-8.9	-9.5 to -11.4	+4.3
Norway	Excluded	+0.7	+0.2	-0.5	+4.1/-3.9	-4.9 to -6.7	See EU-28
	Included	-2.8	-0.6	-1.7	+6.3/-12.4	+1.0 to -0.9	See EU-28
Russia	Excluded	-4.8	+0.8	+1.0	+3.6/-10.4	+1.2 to +2.8	-1.2 to -1.9
	Included	-7.3	-0.2	+1.0	+5.0/-13.0	+6.9 to +5.3	-1.2 to -1.9
Switzerland	Excluded	-0.2	+0.9	-0.7	+1.7/-3.4	-2.4 to -4.0	-4.6
	Included	+0.1	+0.2	-0.5	+3.5/-4.2	-2.6 to -4.2	-4.6
United States	Excluded	+1.3	+0.4	-1.5	+2.4/-4.5	-1.0	-2.3 to -2.8
	Included	+1.7	-0.6	-1.6	+2.9/-5.0	-0.9	-2.3 to -2.8
Average annual rates of change of emissions intensity ($kgCO_2e$ per USD)							
China	Excluded	-5.9	+0.5	-3.9	+5.3/-8.2	-2.9 to -3.9	-4.0 to -4.4
India	Excluded	-4.6	-5.2	-2.8	-0.1/-8.5	-0.3 to -1.1	N/A
Average annual rates of change of total GHG emissions, mitigation goals relative to a BAU baseline ($MtCO_2e$)							
Brazil	Included	+4.2	-0.5	-7.3	+31.8/-18.8	+0.6 to +7.0 (BAU +13.2)	N/A
	Excluded	+2.8	+2.7	+2.5	+0.5/+5.5	+0.6 to +7.0 (BAU +13.2)	N/A
Chile	Excluded	+4.1	+1.9	+2.2%	+10.8/-2.9	+2.3 (BAU +4.6)	N/A
Indonesia	Included	+11.5	N/A	N/A	+18.4/-18.1	N/A	N/A
	Excluded	+7.6	N/A	N/A	+8.8/+4.4	N/A	N/A
Israel	Excluded	N/A	+0.2	+1.1 (2005-11)	+3.7/-1.8	+1.2 (BAU +3.7)	N/A
Korea	Excluded	+5.5	+2.2	+3.0	+9.3/-4.3	-2.8 (BAU +1.6)	-0.2 (BAU +0.8)
Mexico	Excluded	+2.1	+1.7	+2.7 (2005-10)	+4.4/+0.0	-1.0 (BAU +2.6)	+0.5 to +2.8 (BAU +2.1)
South Africa	Excluded	N/A	+2.0	+1.9 (2005-10)	+4.4/-0.5	-1.0 (BAU +3.3)	N/A
Other countries							
Colombia	Excluded	+3.2	+0.3 (2000-04)	N/A	N/A	N/A	N/A
	Included	+2.2	+1.1 (2000-04)	N/A	N/A	N/A	N/A
Costa Rica	Excluded	+6.1	+1.9	N/A	N/A	N/A	N/A
	Included	-0.8	+1.8	N/A	N/A	N/A	N/A
Turkey	Excluded	+4.7	+2.1	+4.2	+9.1/-1.9	N/A	N/A

1. **How to read this table:** This table compares historical rates of change of GHG emissions (or emissions intensity in the case of China and India) with future rates of change needed to meet mitigation targets and goals. For historical data, the annual rates of change are shown for three periods: (i) 1990-2000; (ii) 2000-05; and (iii) 2005-12. The maximum and minimum annual rates of change observed since 1990 (using a three-year rolling average) are also displayed. For the 2020 pledges, the table shows the average annual rate of change needed over the period 2012-20 to meet each pledge. For countries that have submitted a range of pledges, a range of annual rates is included. For INDCs, the annual rate of change needed from 2020 to meet the INDC is shown. These results exclude use of GHG units from market mechanisms. See Annex A for further details on methodology.
2. EU Member States are not included in this table. EU Member States participate in the EU ETS, a binding GHG emissions cap-and-trade system covering nearly half of the EU's emissions. National GHG emissions trends are therefore influenced by the overall EU-wide emissions cap.
3. The following countries have expressed their intention to use carbon credits from market mechanisms to meet part of their targets and goals: Canada, EU-28 (for 30% target for 2020), Iceland, Japan, Korea (for INDC), New Zealand, Norway and Switzerland.

Table 2.5. **Analysis of country mitigation targets and goals** *(cont.)*

4. Norway intends to fulfil its commitment through a collective delivery with the EU and its Member States. In the event that there is no agreement on a collective delivery with the EU, Norway will fulfil the commitment individually. The ambition level will remain the same in this event.
5. China has pledged to reduce its CO_2 intensity and India has pledged to reduce its emissions intensity (excluding agriculture). For the purpose of this exercise, IEA statistics for CO_2 emissions from fossil fuel combustion are used in both cases.
6. Only partial or incomplete time series of national GHG statistics are available for Chile, China, Colombia, Costa Rica, India, Indonesia, Israel, Mexico and South Africa.
7. For Australia, Kyoto Protocol accounting data including LULUCF is used. For other countries with emission reduction targets, UNFCCC data excluding and including LULUCF is used.
8. For Brazil, Chile, Korea, Israel, Mexico and South Africa, national BAU baseline data are used to estimate future rates of change of emissions. Indonesia has yet to publish its national BAU baseline.
9. Rates of change of emissions are influenced by long-term factors (such as population growth or investment in low-carbon RD&D), as well as short-term factors (such as weather conditions or manufacturing output).

Source: Authors' calculations (see Annex A for full list of sources).

Figure 2.2. **Historical rates of change of GDP and emission intensity**

Annual growth rates in 1990-2012

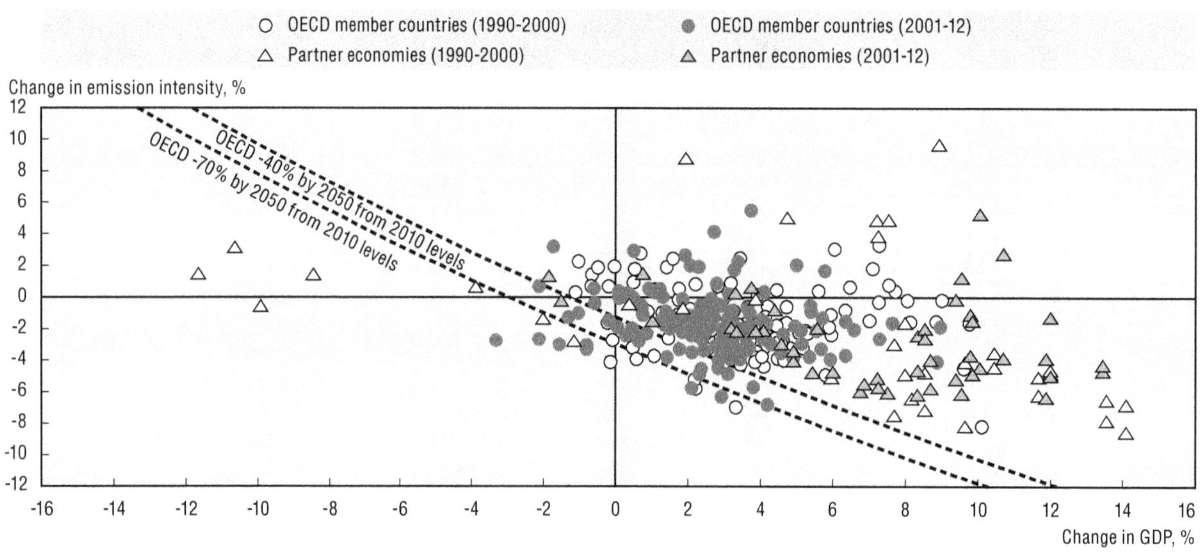

1. **How to read this chart**: This chart shows historical annual rates of change of emissions intensity (excluding LULUCF) plotted against annual GDP growth rates for OECD member countries and partner economies. Each marker represents one country in one year. Markers in the bottom right-hand quadrant show years in which GDP increased while emissions intensity decreased (i.e. decoupling of emissions from GDP growth was achieved). The dashed lines show the average annual changes needed over the period 2010-50 to reduce total GHG emissions from OECD member countries (excluding LULUCF) by 40% and 70% from 2010 levels by 2050.
2. For all countries except China and India, national GHG statistics from UNFCCC inventories were used. IEA statistics on CO_2 from fossil fuel combustion were used for China and India, as their targets are based on CO_2 intensity. For countries with complete time series, annual growth rates were calculated on a three-year rolling average basis to smooth out short-term fluctuations.
3. Emission intensity refers to GHG emissions per unit of GDP (USD 2005 PPP).

Source: Authors calculations based on GHG and GDP data from UNFCCC (2015), Greenhouse Gas Inventory Data (database), and IEA (2015c), CO_2 Emissions from Fossil Fuel Combustion.

StatLink http://dx.doi.org/10.1787/888933272463

Updates on national progress towards targets and goals

A brief overview of each country's progress towards its mitigation targets and goals appears below. Since EU Member States have agreed to jointly fulfil their 2020 and 2030 targets, the EU is treated as a whole.

Australia's GHG emissions would need to decrease by 0.4-3.7% per year in 2013-2020 and 1.6 to 1.9% per year in 2020-30 to meet its 2020 and 2030 emissions reduction targets respectively (including LULUCF and assuming the target is met domestically). Since 1990, strong increases in population and GDP per capita have offset significant decreases in GHG per unit of GDP. Coal continues to dominate the electricity generation mix in Australia, although the carbon intensity of its power sector has fallen since 2006 due to increasing use of gas and renewables. The transport sector is one of the fastest growing sources of GHG emissions. Non-CO_2 emissions, mainly from agriculture, account for around 20% of Australia's total GHG emissions. Livestock populations declined between 2002-10 due to prolonged drought, but have rebounded since 2010, leading to increased methane emissions.

The LULUCF sector has traditionally been the key driver of GHG emissions in **Brazil**. Significantly reduced rates of deforestation since 2004 (albeit from a high baseline level) have resulted in a 41% decrease in total GHG emissions, including LULUCF, between 2005-12. This puts Brazil on track to meeting its goal to reduce GHG emissions by between 36.1% and 38.9% from BAU levels by 2020. Additional measures, however, may be required to counteract the rapid increase in GHG emissions from energy and agriculture, the two sectors that today account for the bulk of emissions. Brazil has a low-carbon energy mix, largely based on the use of renewable energy sources. Renewables, mainly hydro power and biofuels, accounted for more than 40% of total primary energy supply in 2012. However, strong economic growth in the 2000s and the rise of a middle class triggered a rapid increase in energy use, mainly in industry and transport. Rising demand for mobility has led to a doubling of the vehicle fleet, increased energy use and GHG emissions from transport and higher environmental pressures in many urban areas (OECD, 2015).

After strong emissions growth in the 1990s, **Canada's** GHG emissions have plateaued and declined slightly since 2005. Total GHG emissions need to decline by 1.7% per year between 2012-20 to meet its targets to reduce emissions by 17% by 2020 and 30% by 2030 from 2005 levels. Increased GHG emissions compared to 1990 have largely been due to the transport sector and production of synthetic crude oil and bitumen from oil sands, which increased by nearly 450% between 1990-2012; this led to an 82% increase in GHG emissions from the fossil fuel extractive industries sector over the same period (Government of Canada, 2014). The carbon intensity of Canada's power sector is low, at around 150 gCO_2 per kWh in 2012. Indeed, CO_2 emissions from the power sector have declined since 2005 due to less coal use and more use of hydro, solar, wind and nuclear power, together with increased power plant combustion efficiency. In its INDC, Canada announced its intent to develop new regulations to address methane emissions in the oil and gas sector; GHG emissions from natural gas-fired electricity generation; production of chemicals and fertilisers; and HFCs and emissions from heavy-duty vehicles.

Chile's mitigation goal is a 20% reduction in GHG emissions by 2020 from a BAU baseline. Chile's total GHG emissions nearly doubled between 1990-2010. The MAPS Chile project developed and analysed BAU emission scenarios for the country (MAPS Chile, 2015). In the medium GDP growth BAU scenario, GHG emissions continue to rise at +4.6% per year between 2012 and 2020. To achieve the goal, the emission growth rate would need to be reduced to +2.3%. Energy industries and transport sectors account for 40% and 30.5% of Chile's emissions respectively (Government of Chile, 2014). Agriculture accounted for 15% of Chile's total GHG emissions in 2010, compared to the OECD average of 8%. Chile's energy mix relies predominantly on imported fossil fuels (70% of TPES), followed by biofuels and waste. The reliance on coal for electricity generation has been steadily increasing since 2005 and reached 42% in 2013.

China is currently the world's largest GHG emitter in absolute terms, with CO_2 emissions from fuel combustion of over 8.2 $GtCO_2$ in 2012. GHG emissions per capita rose from 3.4 tCO_2e to 8.0 tCO_2e in 1990-2010, although GHG emissions per unit of GDP declined by over 60% over the same period. China's CO_2 intensity decreased by 3.9% per year in 2005-12. Further average annual reductions in CO_2 intensity of 2.9% to 3.9% per year in 2012-20 and 4.0% to 4.4% in 2021-30 are needed to meet its carbon intensity goals for 2020 and 2030. Despite significant investments in non-fossil fuel technologies such as wind, solar and nuclear power, China's energy supply remains dominated by coal. China's 12th Five-Year Plan contains a range of measures and targets that will contribute to climate change mitigation. These include adjustments to the industrial structure and targets for energy efficiency and forest coverage and stock volume.

In 2010, **Colombia's** energy and agriculture sectors were the largest contributors to GHG emissions. Colombia communicated a range of mitigation actions for 2020, including a pledge to increase the share of renewable energy in total installed capacity to at least 77% by 2020. With financial support, Colombia intends to reduce deforestation in the Colombian Amazon rainforest to zero by 2020. It also plans to promote the use of biofuels with the aim of achieving a 20% share of total national fuel consumption by 2020. In 2012, 41% of Colombia's energy supply came from oil, followed by gas and renewable energy, which accounted for about 25% each.

Costa Rica has pledged to implement a "long-term economy-wide transformational effort to enable carbon-neutrality" that will help the country significantly deviate from BAU emission scenarios from now until 2021 and beyond. While Costa Rica's electricity sector already largely relies on renewable energy (92%), its transport sector remains based on fossil fuels. Costa Rica may use CO_2 sinks in the forestry sector to reach its carbon neutrality goal. Previously a net source, Costa Rica's LULUCF sector has been a net CO_2 sink since 1996. Half of Costa Rica's energy supply came from renewable energy, mostly from geothermal sources, in 2012.

The **European Union** has pledged to reduce its GHG emissions by 20% or 30% (conditional) from 1990 levels by 2020. It has translated the 20% target into a commitment under the second commitment period of the Kyoto Protocol (Table 2.1). The European Union has also announced a target to reduce domestic GHG emissions by at least 40% below 1990 levels by 2030. Average annual GHG emission reduction rates of 0.1-1.8% and 1.5-2.8% are needed to meet the 2020 and 2030 targets respectively. Emission reductions in both the ETS and non-ETS sectors are needed to meet the targets. The 2009 Effort Sharing Decision sets national emission targets for 2020. Recent emission reductions have been due to various factors. The economic recession, for example, led to reduced industrial and cement production, as well as reduced rates of passenger and freight transportation. Other factors include policies to increase use of renewable energy sources and energy efficiency. However, CO_2 emissions from the EU-28 are rising again, due in part to an increase in power production from coal in Germany, the United Kingdom and Spain (European Commission, 2014). In addition to its GHG emission reduction target for 2020, the European Union has targets for renewable energy (20% of energy consumption by 2020) and energy efficiency (20% reduction in energy consumption by 2020).

In a joint effort with the European Union, **Iceland** communicated a target of a 30% emission reduction by 2020 compared with 1990. Iceland's total GHG emissions (excluding LULUCF, which is a net source) have decreased by 4% since 1990. Iceland's emission profile is unique as most emissions come from the industry sector, particularly three aluminium smelters and a ferrosilicon plant (OECD, 2014a). Due to Iceland's high share of renewable energy sources in electricity generation, mostly hydro power and geothermal, it has the lowest carbon intensity of electricity generation of the countries studied, at 0.17 gCO_2e per kWh.

India has pledged to reduce its emission intensity (excluding agriculture) by 20-25% from 2005 levels by 2020. Based on data for CO_2 from fossil fuel combustion only, India's emission intensity would need to decrease by 0.3-1.1% per year on average to meet this goal (compared to historical reductions of 2.8% per year in 2005-12). Despite significant reductions in emission intensity since 1990, India's GHG emissions are increasing due to a growing population and rapidly rising levels of GDP per capita coupled with an energy mix that remains dominated by coal. The carbon intensity of the electricity sector is among the highest among OECD member countries and partner economies. According to the latest national data available for 2007 (different from IEA Statistics), agriculture accounted for close to 18% of India's emissions and is not covered by the pledge (Government of India, 2010).

Indonesia's total GHG emissions, excluding LULUCF, grew at an annual rate of between +3.9% and +9.7% per year in 1990-2000. Indonesia has pledged to reduce emissions by 26% or 41% (with support) from BAU levels by 2020. However, a national BAU baseline has yet to be published, which makes an analysis of this pledge challenging. Indonesia's primary energy supply mix remains dominated by fossil fuels and the share of coal is expected to increase from 18.7% to 33% in 2005-25 (Government of Indonesia, 2012). In addition, LULUCF and peat fires are significant sources of CO_2 emissions in Indonesia and accounted for around 74% of total CO_2 emissions in the year 2000, although emissions from LULUCF fluctuate considerably between years (Government of Indonesia, 2012).

Israel has pledged to achieve a 20% reduction in GHG emissions below BAU levels by 2020. In Israel's BAU scenario, GHG emissions excluding LULUCF increase from 78 $MtCO_2e$ in 2011 to 109 $MtCO_2e$ in 2020. Israel's emissions are rising primarily due to the country's relatively high population growth and an increasing standard of living. The energy sector accounts for the largest share of Israel's GHG emissions as the country relies almost entirely on fossil fuels for its energy supply (97% of TPES). Achieving the 20% goal would require Israel to limit the annual increase in its total GHG emissions to +1.2%, compared to +3.7% in the BAU scenario.

After the Great East Japan Earthquake in 2011, **Japan** changed its 2020 commitment from a 25% reduction from 1990 levels to a 3.8% reduction from 2005 levels based on the assumption that no nuclear reactors would restart (Government of Japan, 2013, 2010). Japan's GHG emissions subsequently increased because the use of nuclear power plants was suspended, resulting in more use of thermal power plants (as well as renewable energy sources). However, Japan is reviewing the safety of all 43 reactors not set for decommissioning. This effect outweighed the reduction in emissions from decreased manufacturing output in the wake of the earthquake. Average annual emission reductions of 0.4% are needed to achieve its 3.8% goal.

Korea has pledged to reduce its GHG emissions by 30% from BAU levels by 2020 and 37% below BAU levels by 2030. Average annual emission reductions of 2.8% in 2012-20 and 0.2% in 2020-30 are needed to meet these goals, compared to annual increases of 1.6% and 0.8% per year in the BAU scenarios for 2012-20 and 2020-30 respectively. Korea's GHG emissions decreased significantly following the 1997 East Asian financial crisis, but have been increasing steadily since 2005. While GHG emissions per unit of GDP also decreased after 1997, this trend was reversed after 2008. CO_2 emissions from the transport sector have also increased largely due to a sevenfold increase in the number of passenger cars in Korea between 1990-2012. Energy supply is highly reliant on fossil fuels; as a result, most emissions come from energy use, especially for industries.

Mexico has pledged to reduce its GHG emissions by up to 30% from BAU levels by 2020, and by 25% to 40% (with support) below BAU levels by 2030. For the 2020 pledge, average annual emission reductions of 1.0% would be required, compared to an increase of 2.6% per year in the BAU scenario. For the 2030 pledge, an average annual rate of change of total GHG emissions of +0.5 to +2.8% is needed, compared with +2.1% in the BAU scenario. While Mexico's goal to reduce emissions by 2020 was conditional on international support, its INDC for 2020-30 has an unconditional component. The energy sector, including transport, is the fastest growing source of emissions. However, a shift from oil to gas since 2000 has reduced the carbon intensity of the power sector. Although Mexico has abundant renewable energy resources, the share of renewables in electricity generation actually decreased from 25 to 14% in 1990-2013.

Average annual emission reductions of 3.4-5.5% are needed for **New Zealand** to meet its target to reduce GHG emissions by 5% or 10-20% (conditional) by 2020. To meet its INDC to reduce emissions by 30% from 2005 levels by 2030 (provisional pending LULUCF accounting rules), rates of change of –0.5% to +1.2% in 2020-30 are needed. These emission reduction rates exclude the LULUCF sector. Since LULUCF acts as a significant net sink in New Zealand, lower reduction rates would be needed if removals from LULUCF were taken into account. Most of the growth in emissions since 1990 is due to CH_4 emissions from dairy cattle, CO_2 emissions from road transport and N_2O emissions from agricultural soils (Government of New Zealand, 2013). New Zealand is the only country where agriculture emits more than the energy sector. More than 70% of its electricity comes from renewable energy sources, particularly hydro power.

Norway has pledged to reduce its emissions by 30% or 40% (conditional) from 1990 levels by 2020, and at least 40% by 2030. Average annual emission reduction rates of 4.9-6.7% are needed to meet the 2020 target. Norway intends to deliver the INDC collectively with the EU. Norway intends to use carbon credits from international market mechanisms to meet part of its INDC. The accounting approach for LULUCF will not change the ambition level, since the intention of using LULUCF to meet the commitment is based on accounting for new measures in the sector only. Almost all of the country's electricity already comes from hydro power, although the rest of the energy supply mix is more diversified.

The **Russian Federation's** GHG emissions fell steeply in the 1990s following the collapse of the Soviet Union. However, the trend was reversed after 2000. The country's 2020 target is to reduce GHG emissions by 15% to 25% from 1990 levels. This target permits an increase in emissions of 1.2-2.8% per year in 2012-20. The 2030 target to reduce emissions by 25-30% implies an average annual emission reduction rate of between zero and 1.9%. While the energy intensity of the country's economy has improved since the 2000s, it still remains around double that of most OECD member countries. It could abate significant emissions by replacing old infrastructure, improving energy efficiency and reducing emissions in gas flaring and fugitive emissions from oil and gas production (IEA, 2014). Fugitive emissions from the oil and gas sector amounted to 362 MtCO_2e in 2012 (UNFCCC, 2015d) – greater than the total GHG emissions of Spain.

The average annual rate of change of **South Africa's** GHG emissions would need to be between –3.1% and +0.3% to meet its goal of a 34% reduction below BAU levels by 2020. In the national BAU baseline (known as the "Growth Without Constraints" scenario), GHG emissions increase by 1.0-3.1% per year over the same period. Coal continues to dominate South Africa's energy mix, resulting in a very carbon-intensive power sector. Further, South Africa's economy remains dominated by energy-intensive mining and minerals extraction industries. The energy sector (including power generation and transport) accounted for 78% of total GHG emissions in 2010, up from 75% in 2000 (Letete, 2014).

Switzerland seeks to reduce emissions by 20% or 30% (conditional) from 1990 levels by 2020 and by 50% by 2030. To achieve these targets, average annual emissions must be reduced by 2.4-4.0% and 3.3-4.6% in 2012-20 and 2020-30 respectively. Most of Switzerland's emissions come from the buildings and transport sectors. The carbon intensity of its electricity generation is low due to reliance on nuclear and hydro power. Following the accident at Japan's Fukushima Dai-ichi nuclear power station in 2011, however, Switzerland decided to phase out nuclear power by not replacing its five ageing reactors.

The annual rate of change of **Turkey's** GHG emissions varied between +9.1% to –1.9% in 1990-2012. Turkey's GHG emissions per capita are relatively low, but total emissions are increasing rapidly (OECD, 2014b). Emissions from the energy sector (including power generation and transport) grew by 132% in 1990-2012, making it the fastest growing source of national emissions. Oil, coal and natural gas account for about 90% of total primary energy supply.

While the **United States'** GHG emissions rose steadily during the 1990s and early 2000s, this trend has been reversed since 2005. The United States is on track to reduce emissions by 17% from 2005 levels by 2020. Recent reductions have been due in part to the rapid development of domestic shale gas resources, which has led to a shift from coal to gas in the power sector, as well as the impact of the global financial crisis on economic activity. Broderick and Anderson (2012) estimate this fuel switching accounted for up to half of emission reductions in the US energy sector between 2008-11. Although the carbon intensity of gas is lower than coal, further decreases in the carbon intensity of the power sector will be required to achieve the transition to a low-carbon economy. The transport sector reduced GHG emissions by 9% between 2005-12, partly due to increased fuel efficiency in the vehicle fleet.

References

Broderick, J. and K. Anderson (2012), "Has US shale gas reduced CO_2 emissions? Examining recent changes in emissions from the US power sector and traded fossil fuels", Tyndall Manchester Climate Change Research, *www.tyndall.ac.uk/sites/default/files/broderick_and_anderson_2012_impact_of_shale_gas_on_us_energy_and_emissions.pdf*.

European Commission (2014), *Sixth National Communication and First Biennial Report from the European Union under the UN Framework Convention on Climate Change, http://unfccc.int/files/national_reports/annex_i_natcom_/application/pdf/eu_nc6.pdf*.

European Commission (2009), Directive 2009/28/EC of the European Parliament and of the Council of 23 April 2009 on the promotion of the use of energy from renewable sources and amending and subsequently repealing Directives 2001/77/EC and 2003/30/EC, *http://eur-lex.europa.eu/legal-content/EN/ALL/?uri=CELEX:32009L0028*.

EU (2015), Submission by Latvia and the European Commission on Behalf of the European Union and Its Member States: Intended Nationally Determined Contribution of the EU and its Member States.

Executive Office of the President (2013), *The President's Climate Action Plan*, June 2013, *https://www.whitehouse.gov/sites/default/files/image/president27sclimateactionplan.pdf*.

Finland Ministry of the Environment (2015), "Parliament approves the Climate Change Act", webpage, *www.ym.fi/en-US/Latest_news/Press_release_service/Parliament_approves_the_Climate_Change_A(32988)* (accessed 20 July 2015).

Government of Australia (2014), *Australia's Sixth National Communication to the United Nations Framework Convention on Climate Change*, *http://unfccc.int/national_reports/annex_i_natcom_/items/1095.php*.

Government of British Columbia (2014), *Carbon Neutral BC Annual Report 2014*, *www.2.gov.bc.ca/gov/content/environment/climate-change/reports-data/carbon-neutral-action-reports*.

Government of Canada (2014), *Canada's Sixth National Communication to the United Nations Framework Convention on Climate Change*, *http://unfccc.int/national_reports/annex_i_natcom_/items/1095.php*.

Government of Chile (2014), *Chile's National Greenhouse Gas Inventory, 1990-2010*, *http://unfccc.int/resource/docs/natc/chlnire.pdf*.

Government of India (2012), *India's Second National Communication to the United Nations Framework Convention on Climate Change*, *http://unfccc.int/resource/docs/natc/indnc2.pdf*.

Government of India (2010), *India: Greenhouse Gas Emissions 2007*, Indian Network for Climate Change Assessment, *www.moef.nic.in/downloads/public-information/Report_INCCA.pdf*.

Government of Indonesia (2012), *Indonesia's Second National Communication to the United Nations Framework Convention on Climate Change*, *http://unfccc.int/resource/docs/natc/indnc2.pdf*.

Government of Japan (2013), *Japan's Submission to the United Nations Framework Convention on Climate Change*, *http://unfccc.int/files/focus/mitigation/application/pdf/submission_by_the_government_of_japan.pdf*.

Government of Japan (2010), *"Japan's Submission to the United Nations Framework Convention on Climate Change*, *http://unfccc.int/files/meetings/cop_15/copenhagen_accord/application/pdf/japancphaccord_app1.pdf*.

Government of New Zealand (2013), *New Zealand's Sixth National Communication to the United Nations Framework Convention on Climate Change*, *http://unfccc.int/files/national_reports/annex_i_natcom/submitted_natcom/application/pdf/sixth-national-communication_20131220[1].pdf*.

Government of the United Kingdom (2014), *The United Kingdom's Sixth National Communication to the United Nations Framework Convention on Climate Change*, *http://unfccc.int/national_reports/annex_i_natcom_/items/1095.php*.

IEA (2015a), "World energy balances", *IEA World Energy Statistics and Balances* (database), *http://dx.doi.org/10.1787/data-00512-en* (Accessed 20 July 2015).

IEA (2015b), *World Energy Outlook 2015 Special Report on Energy and Climate*, IEA/OECD Publishing, Paris.

IEA (2015c), "CO_2 emissions by product and flow", *IEA CO_2 Emissions from Fuel Combustion Statistics* (database), *http://dx.doi.org/10.1787/data-00430-en* (accessed 20 July 2015).

IEA (2014), *Energy Policies Beyond IEA Countries: Russia 2014*, IEA/OECD Publishing, Paris, *http://dx.doi.org/10.1787/9789264211506-en*.

IRENA (2015), "Renewable Energy Targets Quadrupled Globally Since 2005, New IRENA Report Finds", Press Release, *www.irena.org/News/Description.aspx?NType=A&mnu=cat&PriMenuID=16&CatID=84&News_ID=413*.

Japan (2015), "Multilateral assessment Japan", presentation at SBI42, Bonn, 4 June 2015, *http://customers.meta-fusion.com/wcm/150601_5044_UNFCCC_SB42_Bonn/download/12_Japan_Presentation_150529_MA_SBI42_4June15.pdf*.

Letete, T. (2014), "South Africa's baseline defined from the long-term mitigation scenarios: Growth without constraints (GWC)", Energy Research Centre, University of Cape Town, *www.oecd.org/env/cc/50025001.pdf*.

MAPS Chile (2015), *Mitigation Options for a Low Carbon Development, Phase 2, Synthesis of Results*, *www.mapsprogramme.org/wp-content/uploads/Chile-Phase-2-Synthesis-of-Results_English_Final.pdf*.

METI (2014), "Strategic energy plan", Ministry of Economy, Trade and Industry, Japan, April 2014, *www.enecho.meti.go.jp/en/category/others/basic_plan/pdf/4th_strategic_energy_plan.pdf.*

National Bureau of Statistics of China (2015), "Statistical communiqué of the People's Republic of China on the 2014 national economic and social development", *www.stats.gov.cn/english/PressRelease/201502/t20150228_687439.html.*

OECD (2015), *OECD Environmental Performance Reviews: Brazil 2014,* OECD Publishing, Paris.

OECD, IEA, ITF, NEA (2015), *Aligning Policies for a Low-carbon Economy*, OECD Publishing, Paris, *http://dx.doi.org/10.1787/9789264233294-en.*

OECD (2014a), *OECD Environmental Performance Reviews: Iceland 2014*, OECD Publishing, Paris, *http://dx.doi.org/10.1787/9789264214200-en.*

OECD (2014b), *OECD Economic Surveys: Turkey 2014*, OECD Publishing, Paris, *http://dx.doi.org/10.1787/eco_surveys-tur-2014-en.*

UNFCCC (2015a), *Report of the Conference of the Parties on its Twentieth Session*, Lima 1-14 December 2014, *http://unfccc.int/resource/docs/2014/cop20/eng/10a01.pdf.*

UNFCCC (2015b), "INDCs as communicated by Parties", web portal, United Nations Framework Convention on Climate Change, *www.4.unfccc.int/submissions/indc/Submission%20Pages/submissions.aspx* (accessed 20 August 2015).

UNFCCC (2015c), Kyoto Protocol website, United Nations Framework Convention on Climate Change, *http://unfccc.int/kyoto_protocol/items/2830.php* (accessed 06 July 2015).

UNFCCC (2015d), *Greenhouse Gas Inventory Data* (database), United Nations Framework Convention on Climate Change, *http://unfccc.int/ghg_data/items/3800.php* (accessed 20 June 2015).

UNFCCC (2011a), "Compilation of economy-wide emission reduction targets to be implemented by Parties included in Annex I to the Convention", United Nations Framework Convention on Climate Change, *http://unfccc.int/resource/docs/2011/sb/eng/inf01r01.pdf.*

UNFCCC (2011b), "Compilation of information on nationally appropriate mitigation actions to be implemented by Parties not included in Annex I to the Convention", United Nations Framework Convention on Climate Change, *http://unfccc.int/resource/docs/2011/awglca14/eng/inf01.pdf.*

REN21 (2015), "Renewable energy interactive map", Renewable Energy Network for the 21st Century, Paris.

White House (2015) "Fact sheet: Administration takes steps forward on climate action plan by announcing actions to cut methane emissions", *https://www.whitehouse.gov/the-press-office/2015/01/14/fact-sheet-administration-takes-steps-forward-climate-action-plan-anno-1.*

Xinhuanet (2014), "China unveils energy strategy, targets for 2020", 19 November 2014, *http://news.xinhuanet.com/english/china/2014-11/19/c_133801014.htm.*

Climate Change Mitigation
Policies and Progress

Chapter 3

Carbon pricing

This chapter discusses key trends in the use of carbon pricing instruments. It outlines developments in the use of energy and carbon taxation and emissions trading systems to reduce greenhouse gas emissions. It considers the effective tax rates implied by energy and carbon taxes in different countries and in different sectors. For emissions trading systems, recent developments and different approaches to coverage and allocation of emission allowances are highlighted. This chapter also examines budgetary support and tax expenditures for the production and consumption of fossil fuels.

Carbon pricing instruments

An increasing number of countries are implementing explicit carbon and energy taxes, emissions trading systems or a combination of both. These policies aim to influence behaviours and investment patterns by placing an explicit price on greenhouse gas (GHG) emissions. The policies can be complementary and different pricing instruments may be appropriate to address GHG emissions from different sources. The World Bank (2015) estimates the combined value of carbon pricing mechanisms globally in 2015 to be just under USD 50 billion.

Table 3.1 summarises carbon pricing instruments used in the countries studied. Many countries have a mix of carbon and energy taxes and emissions trading systems (ETSs) at regional, national and sub-national scales. The effectiveness of carbon pricing instruments can be increased by minimising overlap between them and other policy instruments in different sectors and at different levels of governance.

Table 3.1. **Carbon pricing instruments**

Country	Location, instrument and year	Coverage	Price per tCO_2e
Canada	Quebec ETS (2013)	Sub-national: power generation, industry (~30% total emissions)	USD ~10 in 2014
	British Columbia carbon tax (2008)	Sub-national: fossil fuels used by businesses (~70% total emissions)	CAD 30 (USD 24, 2014)
	Alberta carbon tax (2007)	Sub-national: industry, oil and gas (~45% total emissions)	CAD 15 (USD 12) in 2015; CAD 30 (USD 24) by 2017
Chile	Chile carbon tax (planned for 2017)	National: power plants, industry (~55% total emissions)	USD 5 in 2017
China	Beijing ETS (2013)	Sub-national: power generation, industry, buildings (~50% total emissions)	USD ~8 in 2015
	Tianjin ETS (2013)	Sub-national: power generation, industry, buildings (~60% total emissions)	USD ~4 in 2015
	Hubei ETS (2014)	Sub-national: power generation, industry (~35% total emissions)	USD ~4 in 2015
	Shanghai ETS (2013)	Sub-national: industry, aviation, transport, buildings (~50% total emissions)	USD ~5 in 2015
	Shenzhen ETS (2013)	Sub-national: power generation, industry, buildings (~38% total emissions)	USD ~7 in 2015
	Guangdong ETS (2013)	Sub-national: power generation, industry (~42% total emissions)	USD ~5 in 2015
	Chongqing ETS (2014)	Sub-national: power generation (~38% total emissions)	USD ~4 in 2015
	China ETS (planned for 2016)	National: coverage TBC	-
Costa Rica	Tax on fossil fuels (1997)	National: fossil fuels	3.5% tax
Denmark	Tax on mineral oil products and certain energy products (1991)	National: fuels, electricity (~45% total emissions, excludes EU ETS sectors)	USD 31 in 2014
EU-28	EU ETS (2008)	Supranational: power generation, industry, aviation (~45% total emissions)	EUR ~7 (USD ~8) in 2015
Finland	Fossil fuel tax (1990)	National: heating and transport fuels (~15% total emissions)	EUR 35-60 (USD 39-67)
France	*Contribution climat-énergie* (climate-energy contribution, 2014)	National: petroleum products, natural gas, coal (exemptions for industries covered by EU ETS)	EUR 7 in 2014, EUR 14.5 in 2015, EUR 22 in 2016
Iceland	Carbon tax (2010)	National: liquid fossil fuels and electricity consumption (exemptions for industries covered by EU ETS)	USD 10
India	Coal cess (2010)	National: imported and domestic coal	INR 200 (USD 3) per tonne of coal
Ireland	Carbon tax (2010)	National: fuels for transport, natural gas, coal and peat (exemptions for industries covered by EU ETS)	EUR 20 (USD 22)
Japan	Tax for climate change mitigation (2012)	National: coal, oil, natural gas (~90% total emissions)	JPY 192 (USD 1.6) in 2014, JPY 289 in 2016 (USD 2.4)
	Tokyo ETS (2010)	Sub-national: commercial and industrial buildings (~20% total emissions)	JPY ~4 600 (USD ~38) in 2015
	Saitama ETS (2011)	Sub-national: buildings (~16% total emissions)	

Table 3.1. **Carbon pricing instruments** *(Cont.)*

Country	Location, instrument and year	Coverage	Price per tCO_2e
Korea	Korea ETS (2015)	National: power generation, industry, waste, aviation (~66% total emissions)	KRW 9 610, (USD 9) in 2015
Mexico	Carbon tax (2012)	National: fossil fuel sales and imports (excluding natural gas)	MXN 10-50 (USD 0.6-1.8)
New Zealand	NZ ETS (2008)	National: forestry, power generation, fuels, waste, industry (~50% total emissions)	NZD ~7 (USD ~5) in 2015
Norway	Carbon tax (1991)	National: mineral products, natural gas and LPG (~50% total emissions, exemptions for industries covered by EU ETS, except offshore petroleum industry)	NOK 25-419 (USD 3-52)
Portugal	Green tax reform (2014)	National: fossil fuels (~26% total emissions, exemptions for industries covered by EU ETS)	EUR 5 (USD 5.6)
Slovenia	Carbon tax (1997)	National: coal, lignite, LPG, natural gas, fuel oils, landfills	EUR 0.0144 (USD 0.016)
South Africa	Carbon tax (planned for 2016)	National: fossil fuels for heating and motor fuels (~25% total emissions)	ZAR 120 (USD 9.6) in 2016
Sweden	Sweden carbon tax (1991)	National: fossil fuels, exemptions for electricity production and industry	SEK 1 050 (USD 130) in 2015
Switzerland	Swiss ETS (mandatory since 2013)	National: industry (~10% total emissions)	CHF ~9.5 (USD ~9) in 2015
	Carbon tax (2008)	National: heat and process fuels (~30% total emissions, exemptions for energy-intensive industries with mandatory emissions reductions commitment or participation at Swiss ETS)	CHF 60 (USD 64) in 2015
United States	California ETS (2013)	Sub-national: power generation, industry (~35% total emissions)	USD ~13 in 2015
	Regional Greenhouse Gas Initiative (RGGI) (2009)	Sub-national: power generation (~20% total emissions)	USD ~6 in 2015

Sources: Republic of South Africa (2013); Ecofys/World Bank (2014); Ecologic Institute and eclareon (2014); Government of British Columbia (2014); Korea (2014); Ministry of Sustainable Development and Energy of France (2014); Swiss Federal Office for the Environment (2014); New Zealand (2015); OECD (2015a, 2013a); Swedish Tax Agency (2015); World Bank (n.d.).

Carbon and energy taxation

Environmentally-related taxation can improve environmental quality by using price signals to shift investment and behaviour patterns. In most OECD member countries, taxes on energy products account for the greatest share of environmentally-related tax revenues (more than 60% in most countries), followed by taxes on motor vehicles. Among partner economies, nearly 90% of revenues from environmental taxes in South Africa and Colombia stem from energy products, while in the People's Republic of China (hereafter "China"), energy products account for 48% of revenues and motor vehicles for 35% (OECD, 2015a).

OECD member countries and partner economies widely use energy taxes to price negative externalities and tackle both local air pollution and climate change. Energy taxes affect the prices of energy and consequently influence the amount and type of energy used. Countries with higher per capita GDP tend to tax energy use at higher effective rates. For example, the 2003 EU Energy Tax Directive significantly shapes energy tax policy in the European Union. Besides taxes, other measures, such as differential value added tax (VAT) rates on energy products, also affect relative prices of energy and, consequently, the use of energy. The OECD analysed energy taxation in 41 countries, including OECD member countries, Argentina, Brazil, China, India, Indonesia, the Russian Federation and South Africa. Of these, 18 countries apply a differential VAT rate to energy products. Although countries tax energy use differently, countries reviewed in the OECD report *Taxing Energy Use* (OECD, 2015b, 2013a) show similar patterns of taxation.

Energy for transport is taxed more highly than for heating and process or electricity generation in every country studied except Brazil. Within transport use, oil products for road use, particularly gasoline and diesel, are taxed more highly than other forms of energy use in all countries except in the United States. A higher tax rate on gasoline effectively provides a tax preference to diesel. In the heating and process category, oil products are taxed at the highest rates compared with other fuels in most countries. Over 85% of coal used for heating and process in the 41 countries is left untaxed, despite its higher carbon intensity and the negative impacts coal combustion has on local air quality. Energy used to generate electricity is also taxed at lower rates, when compared on a CO_2 emissions basis, than in transport across countries (Figure 3.1).

Figure 3.1. **Graphical profile of energy use and taxation across all carbon emissions from energy use**

In OECD member countries and seven partner economies

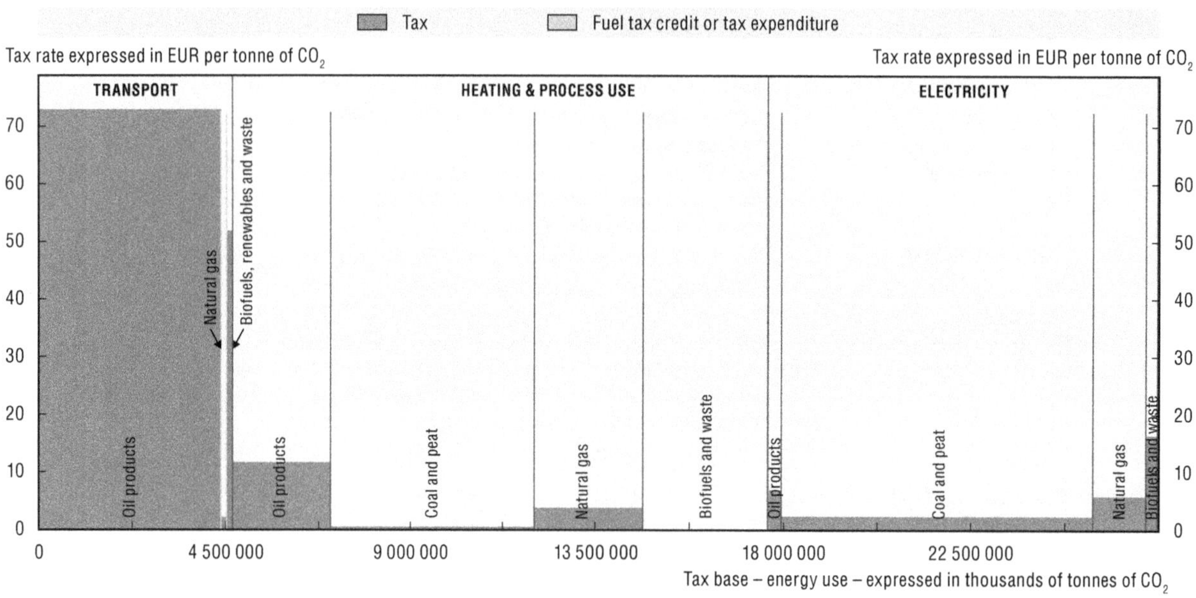

1. Tax rates are as of 1 April 2012 (except 1 July 2012 for Australia and Brazil and 4 April 2012 for South Africa); energy use data are for 2009 from IEA (2011). Figures for Canada, India and the United States include only federal taxes.
2. The 41 countries include OECD member countries, Argentina (which is not reviewed in this report), Brazil, China, India, Indonesia, the Russian Federation and South Africa. Costa Rica, Colombia, Latvia and Lithuania are not reviewed in OECD 2015b or 2013a.

Source: OECD (2015b) *Taxing Energy Use 2015: OECD and Selected Partner Economies*; *Taxing Energy Use – A Graphical Analysis* (OECD, 2013a) for all other countries.

Current policies and taxation rates are sending uneven price signals to consumers, producers and investors across and within countries and sectors. There is scope to improve the use of taxation as a policy tool to help reduce GHG emissions from the energy sector. Fuels such as kerosene, diesel and fuel oil do not have the same energy content and emission characteristics. Therefore, tax rates that are equal in physical terms or in energy terms will not be equal in carbon terms and vice versa. Harmonising a tax rate across different fuels on a carbon content basis would increase tax rates per unit of energy on more carbon-intensive fuels (OECD, 2013a).

Carbon taxes are taxes on energy products designed to reflect the CO_2 emission intensity of different energy sources. Explicit carbon taxes are often introduced at a low level and gradually increased over time. In **Sweden**, a green tax reform in the early 2000s increased the level of carbon tax, resulting in a tax rate of EUR 119 per tCO_2 in 2013, although there are significant exemptions; this was significantly above the EU ETS allowance price (Ecofys/World Bank, 2014; OECD, 2014a). In **Portugal**, the new carbon tax rate is indexed to the previous year's average EU ETS allowance price, using allowance auction data. In **Switzerland**, the initial level of the tax was CHF 12 per tCO_2, which was subsequently raised each year according to a predefined schedule; in 2014, the tax rate was raised significantly (from CHF 36 to CHF 60 per tCO_2) because interim emissions targets had not been met. In 2012, **Japan** introduced a "Tax for Climate Change Mitigation" to reduce CO_2 emissions from energy sources, which account for about 90% of total emissions. The tax rate will increase to JPY 289 per tCO_2 over 3.5 years.

In some cases, a share of the revenues from CO_2 taxes has been allocated to support low-carbon technologies. Switzerland intends to allocate revenues from its carbon levy to reducing emissions from buildings (but no more than CHF 300 million/year) and a clean technology fund (CHF 25 million per year). Japan's climate change tax revenues are estimated at JPY 39 billion in 2012 and JPY 260 billion for each year after 2016; they are being used to reduce energy-related CO_2 emissions through innovation, energy saving and promotion of renewable energy (Ecofys/World Bank, 2014; OECD, 2015f).

Some countries have taken steps to minimise the overlap of carbon and energy taxes with other climate policy instruments. For example, some EU members have established carbon taxes in the energy and industry sectors, but have granted reduced rates or exemptions to installations covered by the EU ETS. **Norway** exempts the sectors covered by the EU ETS from its carbon tax (except from the offshore petroleum sector, which is both subject to the EU ETS and the highest rate of the carbon tax). **Denmark** has also exempted some industries that participate in the EU ETS from its carbon tax.

Other examples of tax reforms to price carbon include the following:

- **British Columbia** in Canada launched a revenue-neutral tax reform in 2008 covering over 70% of the province's GHG emissions. The tax, which applies to all combustion sources of all fossil fuels, was introduced at a rate of CAD 10 per tonne of CO_2. Annual increases of CAD 5 per tonne of CO_2 followed until the tax reached CAD 30 per tonne of CO_2 in 2012. Tax revenues were fully recycled via a combination of corporate and income tax cuts, phased in over time (Harrison, 2013). Following a review of the carbon tax in 2014, including its revenue neutrality and impact on businesses, the provincial government decided to maintain the tax rate at its current level and not to expand the carbon tax base to industrial processes (Government of British Columbia, 2014).
- **Mexico** introduced a new tax on the sale of various fossil fuels in 2014, partly based on their carbon content relative to that of natural gas. The tax rates have been set to USD 5.91 cents per litre (L) for propane, USD 10.38 cents per L for gasoline, USD 12.59 cents per L for diesel and USD 27.54 per tonne for coal (anthracite). Natural gas is exempted from the tax. Companies covered by the tax can meet their obligation by purchasing an equivalent number of credits from projects in Mexico funded by the Clean Development Mechanism. To gain passage through the Congress, the proposed tax rates had to be significantly modified; for example, the tax rate finally imposed on oil coke was reduced by 92% and that on coal by 85% (SEMARNAT, 2014).
- In January 2015, as part of its green fiscal reform, **Portugal** introduced a carbon tax that is expected to raise EUR 95 million. Revenue from this tax is to be expended on encouraging cleaner and greener behaviour in the transport sector (electric cars, bike sharing and car sharing) and support sustainable forest and biodiversity. The carbon tax applies to sectors not covered by the EU ETS (Moreira da Silva, 2015).
- **India** established a coal tax as part of its 2010/11 budget. Part of the revenue from the coal tax is intended to finance a National Clean Energy Fund for research and development and innovative clean energy projects. The tax was introduced on imported and domestically-mined coal at a level of INR 50 per metric tonne. In July 2014, the Indian government increased the tax to INR 100 (EUR 1.43). In February 2015, the government

proposed a further increase in the tax rate to INR 200 (EUR 2.9) and to use the revenues to support renewable energy projects (The Times of India, 2015).

Several other countries plan new carbon taxes. South America's first carbon tax was approved in **Chile** for 2017. Applying to large factories and the electricity sector, the carbon tax should cover around 55% of Chile's national emissions. The intended price of USD 5 per tonne of CO_2 is lower than that of the EU ETS (Galbraith, 2014). **South Africa** postponed the implementation of its carbon tax to 2016. The proposed tax rate starts at ZAR 120 (USD 10) in 2016 and rises at 10% per annum until 2019, although the final tax rate has yet to be confirmed (OECD, 2015g).

Emissions trading systems

Cap-and-trade emissions trading systems (ETSs) set a cap on GHG emissions or intensity, with tradable emission allowances allocated to emitters. Emissions trading systems can reduce GHG emissions cost effectively by providing incentives for emissions abatement where it can be done at least cost. The cap level significantly influences the carbon price in an ETS; setting the cap too high results in a low carbon price that provides little incentive to invest in carbon abatement.

Under an ETS, when the cap is binding, adding other mitigation policy instruments in ETS-covered sectors will not abate additional emissions in those sectors as long as the overall cap remains unchanged (OECD, 2011). Nonetheless, policy instruments often have multiple objectives (including climate and non-climate policy objectives) and can help to raise awareness, share information and support RD&D and technological innovation; these are needed to lower abatement costs and achieve deeper emission reductions in the long term. Multiple policy instruments should complement rather than substitute each other. Examples of complementary instruments are information and labelling policies for the energy efficiency of household appliances, which can improve the effectiveness of carbon pricing policies in the energy sector. The cost effectiveness of individual policies, as well as policy mixes as a whole, needs to be regularly monitored and assessed.

Progress is being made on implementing ETSs at the international, national and sub-national level in the countries studied. Twenty-five of the countries studied participate in the **EU ETS** (including Iceland and Norway, which are linked to the EU ETS). Other existing systems include the **New Zealand ETS**, the **Regional Greenhouse Gas Initiative (RGGI)** in the north-eastern states of the United States and the **Tokyo** and **Saitama ETSs** in Japan. Following its launch in 2008, **Switzerland's** ETS became mandatory in 2013, while **California** and **Quebec** formally linked their systems in 2014. **China**, which established pilot systems in seven cities and provinces, is planning to launch a national ETS in 2017. In 2015, **Korea** launched an ETS covering over 500 companies in the power, steel, cement and manufacturing industries. Further national systems are under consideration in **Brazil, Chile, Mexico, Turkey** and the **Russian Federation**, while further sub-national systems are planned in the state of **Washington**, as well as in **Ontario and Manitoba**. Ontario, Canada's second-biggest producer of greenhouse gases, has announced it will implement a cap-and-trade system and integrate it with that of California and Quebec (Government of Ontario, 2015). California, British Columbia, Manitoba, Ontario and Quebec are members of the Western Climate Initiative (WCI), which aims to establish a regional cap-and-trade system.

Recent developments in the EU ETS

The EU ETS, the world's largest GHG emissions trading system, remains the EU's flagship policy for meeting its climate and energy goals. It operates in the 28 EU Member States, as well as in Norway, Iceland and Liechtenstein. Under the EU ETS Directive (2003/87/EC), the system covers over 11 000 installations in the power generation and manufacturing sectors, as well as commercial aviation (since 2012). Together, these installations account for just under half of the EU's total GHG emissions, although there is significant variation between countries (Figure 3.2).

Figure 3.2. **Share of national GHG emissions covered by the EU emissions trading system**

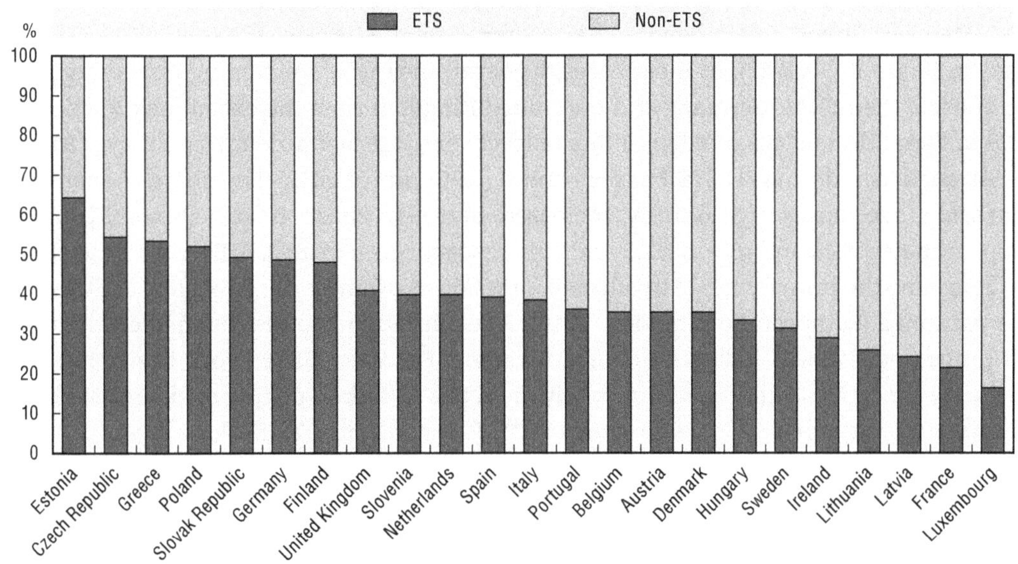

Source: EEA (2015), EU Emissions Trading System (ETS) Data Viewer, *www.eea.europa.eu/data-and-maps/data/data-viewers/emissions-trading-viewer* (accessed 10 June 2015).

StatLink http://dx.doi.org/10.1787/888933272470

The EU ETS is in its third phase (2013-20). The main changes between the second and third phase were: (i) a transition from a system of national caps to a single, EU-wide cap that will decrease by 1.74% per year until 2020 (due to the decreasing cap, emissions in 2020 will be 21% lower than in 2005); (ii) an increase in the share of auctioned allowances (from less than 4% in Phase II to more than 40% in Phase III); (iii) harmonised allocation rules based on performance benchmarks for the remaining free allocation of allowances; and (iv) expansion of the scope to cover CO_2 emissions from petrochemicals, ammonia and aluminium, N_2O emissions from the production of nitric, adipic and glyocalic acid, and PFC emissions from the production of aluminium.

A surplus of emissions allowances has built up in the system, largely from lower-than-anticipated emissions levels. Lower levels are due to the economic crisis combined with high levels of supply of international credits from the Clean Development Mechanism. This has weakened the carbon price signal, which lowers the incentive for covered installations to invest in low-carbon technologies. Carbon emission allowances trade at about EUR 7 per tonne of CO_2e (Henbest, 2015). The European Commission has postponed the auctioning of 900 million allowances (known as "back-loading") to reduce the surplus in the short term; it is also considering longer-term structural reforms to address the issue. The European

Commission has proposed accelerating the annual reduction of the emissions cap from 1.74% to 2.2%. In addition, a proposed market stability reserve from 2021 would improve the system's resilience to major shocks, as well as address built-up emission allowances.

Emission allowances are allocated to ETS participants through different approaches. During the first and second phases, most European Emissions Allowances (EUAs) were allocated for free (1 EUA equals 1 $MtCO_2e$). Free allocation is based on performance benchmarks; installations that meet the benchmark receive a greater share of free allowances. The benchmarks, established on the principle of "one product = one benchmark", generally reflect the average performance of the 10% best-performing installations in the European Union producing that product. In sectors such as manufacturing, the share of freely allocated EUAs remains as high as 80% (in 2013).

Member States could auction 5% of their total EUAs in Phase I (2005-07) and 10% in Phase II (2008-12). In most cases, however, they used auctioning only marginally to allocate their EUAs. **Germany** and the **United Kingdom** were the exceptions, auctioning 209 million EUA and 122.8 million EUAs respectively, together accounting for over 81% of EUAs auctioned during EU ETS Phases I and II (CDC Climat, 2013). The share of auctioned emission allowances significantly increased between Phase II and Phase III, up to approximately 52% and 40% in 2013 and 2014 respectively (Figure 3.3). Eventually, auctions will become the primary mode to allocate emission allowances. By 2020, 70% of EUAs will be auctioned (European Commission, 2015). Free allocation is often justified on the basis of protecting national industrial competitiveness. However, OECD work has found that industry concerns about loss of competitiveness due to carbon pricing measures are often over-stated (Lanzi et al., 2013; Arlinghaus, 2015; Flues and Lutz, 2015).

Figure 3.3. **Allocation method used for emission allowances in the EU emissions trading system**

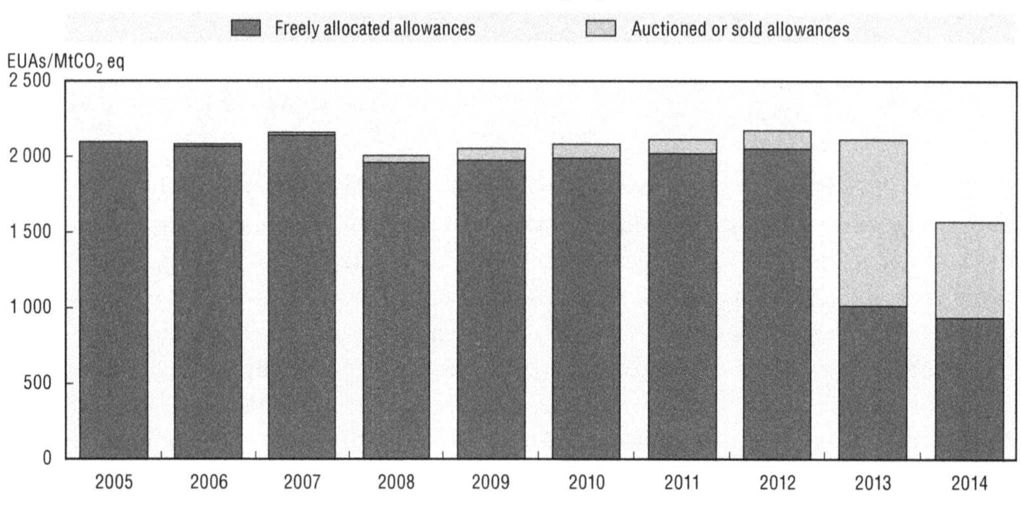

Source: EEA (2015), EU Emissions Trading System (ETS) Data Viewer, *www.eea.europa.eu/data-and-maps/data/data-viewers/emissions-trading-viewer* (accessed 10 June 2015).

StatLink http://dx.doi.org/10.1787/888933272483

Recent developments in other emissions trading systems

China has been experimenting with **seven regional pilot emissions trading systems that cover** five cities (Beijing, Tianjin, Chongqing, Shanghai and Shenzhen) and two provinces (Hubei and Guangdong). These seven areas account for 18% of China's population and

28% of its national GDP (Shen, 2013). Building on these pilot experiences, China is designing a three-year pilot national ETS, which will be launched in 2017. It is initially expected to cover CO_2 emissions from the power generation, metallurgy and nonferrous metals, building materials, chemicals and aviation sectors, together accounting for around 3 000-4 000 $MtCO_2$ (ICAP, 2015a; OECD, 2015c).

The **Korea** ETS, launched in January 2015, set its cap at 573 $MtCO_2e$ in 2015. It covers about two-thirds of the country's total emissions and applies to 525 business entities. During Phase I (2015-17), all allowances are allocated for free, either according to the average GHG emissions of the base year (2011-13) or to benchmarks based on previous activity data. The allocation of free allowances will decrease over time: from 100% free allocation in Phase I to 97% in the Phase II (2018-20), to less than 90% in Phase III (2021-25) (ICAP, 2015b).

In July 2014, the **Australian** government delivered on its election commitment to repeal a carbon pricing mechanism established in 2012 as part of a Clean Energy Future Plan. By way of replacement, the government has introduced a suite of new measures, including an Emissions Reduction Fund that purchases emissions abatement via a reverse auction (OECD, 2014b). The first auction under the fund was held in April 2015, following which the government contracted to purchase over 47 million tonnes of emission reductions. A second auction has been scheduled for November 2015.

Allocation of allowances and coverage

In China, the seven ETS pilots use different approaches to allocation. However, most of these ETSs combine free allocation based on benchmarks with grandfathering based on historical emissions or emission intensity (Zhou, 2015). Around 80% of allowances in the California ETS are auctioned, with the remaining allowances allocated for free based on benchmarks. Most allowances in the Regional Greenhouse Gas Initiative (RGGI) are auctioned (IETA, 2015). In the New Zealand ETS, most allowances are allocated for free based on performance data from the previous year, although the share of free allowances is expected to decline.

Several countries have increased the coverage of their ETSs over time. This increases the number of sources covered by a uniform carbon price, as well as opportunities to exploit cost-effective emission reductions. The coverage of **China's** pilot ETSs varies – some include service industries, heat production, transport, buildings, oil and gas production, and automobile production facilities in addition to the power sector, steel, cement and other manufacturing industries (IETA, 2013). The **New Zealand** ETS is the only national ETS to cover forestry activities (which can earn tradable carbon credits and can incur a requirement to purchase credits if there is felling) in addition to power generation, industrial processes and liquid fossil fuel production facilities. The agriculture sector is not explicitly covered by any ETSs to date, although the NZ ETS features mandatory reporting requirements for biological emissions from agriculture and waste (EDF/IETA, 2014).

Linking emissions trading systems means that allowances from one ETS can be traded and used for compliance with others. As with increasing the coverage of an ETS, linking two or more ETSs together increases the size of the pool of low-cost abatement options that covered entities can access. Clearly, this requires compatibility; in particular, the caps of all ETSs concerned need to be set below BAU levels; otherwise, oversupply in one ETS will collapse prices. A bottom-up network of linked ETSs (including bilateral linking agreements) is emerging. Linkages were established between the California Cap-and-Trade Programme and the Quebec Cap-and-Trade System in 2014, and a linkage is being negotiated between the EU ETS and Switzerland's ETS.

Support for fossil fuels

Many OECD member countries and partner economies support the production of fossil fuels through direct transfers or preferential tax treatment. The consumption of fossil fuels can also be supported through price controls regulating the cost of energy to consumers, direct transfers, consumer rebates on energy-product purchases and targeted tax relief.

A number of international initiatives in recent years have called for the reform of harmful fossil-fuel subsidies. In 2009, for example, G-20 leaders committed to rationalise and phase out over the medium term inefficient fossil-fuel subsidies that encourage wasteful consumption (G-20, 2009). The Asia-Pacific Economic Co-operation (APEC) member economies made a similar pledge, while committing to provide those in need with essential energy services (APEC, 2009). APEC and G-20 economies have since undertaken self-reporting of their fossil-fuel subsidies, although lack of common definitions limits the comparability of these reports. Some G-20 economies have agreed to reciprocal peer reviews of fossil-fuel subsidies, with the United States and China volunteering in 2014 to be first. Further, a Friends of Fossil-Fuel Subsidy Reform initiative has brought together like-minded countries beyond the G-20 to advocate reform (Government of New Zealand, 2015). The OECD has contributed to these various initiatives by sharing its expertise and facilitating the exchange of relevant information among its member countries and other interested parties (OECD, 2015d).

At the regional level, the first set of recommendations adopted in 2013 by the European Commission's European Resource Efficiency Platform (EREP) stated the EU and its Member States should urgently phase out environmentally harmful subsidies. It put special emphasis on subsidies to fossil fuels and the use of water in agriculture, energy and industry (European Commission, 2014). Some regional development banks have also taken steps to evaluate or reform fossil-fuel subsidies in the countries in which they operate through technical cooperation. The Asian Development Bank, for example, supports monitoring and evaluating of fossil-fuel subsidies in some member countries, while the Inter-American Development Bank assists with analysing subsidies for the production or use of fossil fuels in Latin America and the Caribbean (ADB, 2011; IADB 2013).

The IEA and the OECD also both analyse subsidies and other measures supporting fossil fuels. The IEA calculates the price gap between local fuel prices and a set of reference prices to estimate the extent to which fossil fuels are under-priced in various countries. Using this method, the IEA estimates that fossil-fuel subsidies in 40 countries totalled USD 548 billion in 2013 (IEA, 2014). These estimates are particularly well suited for analytical work at the macroeconomic level. However, the price-gap method does not capture some forms of support for the production and consumption of fossil fuels, particularly in OECD member countries.

In 2010, the OECD started to collect complementary information on all budgetary transfers and tax expenditures that encourage the production or consumption of fossil fuels in its member countries. This has become part of a regular OECD exercise; the 2015 inventory was recently expanded to cover almost 800 policies in OECD member countries and six partner economies (Brazil, China, India, Indonesia, the Russian Federation and South Africa). The measures identified in the 2015 inventory had a value of USD 160-200 billion annually over 2010-14, with support for consumption of petroleum products accounting for the bulk of that amount (OECD, 2015d). Compared with the previous edition of the inventory (OECD, 2013b), which focused on OECD member countries only, the total level of support is now decreasing after having peaked twice in 2008 and 2011/12, although the level of total consumer support varies significantly between countries (Figure 3.4). The decline in total support is partly due to lower international oil prices, as well as policy reforms; these reforms signal an intention on the part of many governments to depart from earlier practices and move towards more fiscally and environmentally sustainable growth patterns (OECD, 2015d).

Figure 3.4. **Total consumer support for fossil fuels**

Expressed as a share of the revenues of the energy-related component of environmentally related taxes, average for 2010-12

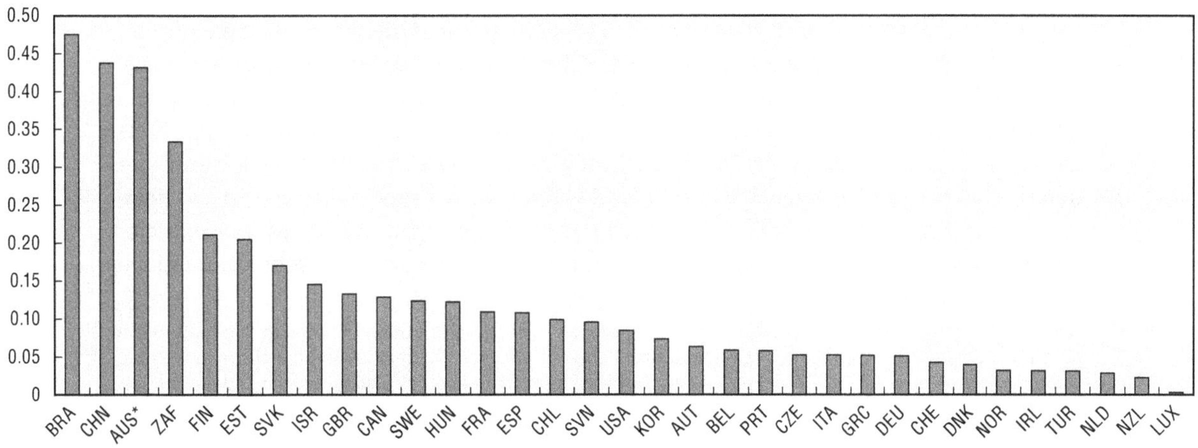

Notes: * The data for Australia include the country's large fuel tax credits, which alone explain the relatively high ratio observed for that particular country. This measure serves to rebate some of the excise taxes that businesses pay on their purchases of fuel there. Data for Brazil and Greece are for the period 2010-11 only.

Data on the revenues countries derive from environmentally-related taxes (including taxes related to the use of energy, motor-vehicle taxes, and other environmental fees and levies (e.g. on waste and water use) are regularly collected by the OECD and made available through the Organisation's database of instruments used for environmental policy (*www.2.oecd.org/ecoinst/queries/*).

Source: OECD (2015d), OECD Companion to the Inventory of Support Measures for Fossil Fuels 2015, OECD Publishing, Paris.

StatLink *http://dx.doi.org/10.1787/888933272496*

The **Indonesian** government reformed its petrol and diesel price-setting regime in January 2015. Until 2014, fuel and electricity subsidies accounted for over 20% of total government spending. The Indonesian government subsequently phased out gasoline subsidies entirely in its revised 2015 budget, leaving in place only smaller subsidies for liquefied petroleum gas (LPG), diesel fuel and kerosene. Subsidies for diesel were capped at IDF 1 000 (USD 0.08) per litre. Indonesia retained the small subsidy on diesel because of its use in public and freight transport. Fuel subsidies were earmarked in the 2015 budget to make up more than 13% of total government expenditure, but this has now been decreased to 1% – equivalent to a USD 14 billion decrease in a single year. Consumer subsidies in Indonesia disproportionately benefited rich households prior to their recent reform. Since the removal of subsidies could harm the poorest segments of the population, the social impacts of such reforms need to be addressed (Box 3.1) (OECD, 2015d, 2015e).

In **India**, energy prices have traditionally been highly regulated, with prices often kept below cost. Gasoline prices were partially deregulated in June 2010, and then fully deregulated in October 2014 during a period of low international gas prices. In 2012, the Indian government reduced consumer subsidies for diesel fuel, resulting in savings of

Box 3.1. **Distributional impacts of fossil fuel subsidy reform**

By increasing energy prices, some mitigation policies risk impeding access to modern energy and affecting disposable household income. Complementary policies are therefore important (IPCC, 2014). According to an OECD modelling analysis, phasing out subsidies for consumption of fossil fuels in Indonesia could help decrease energy-related CO_2 emissions by 10-12% by 2020. It could also result in welfare gains if a redistribution scheme, such as cash transfers, is put in place to make the fossil fuel subsidy reform progressive and pro-poor (Durand-Lasserve et al., 2015).

In parallel, new OECD work challenges the idea that wider use of energy taxes is regressive and hits the poor harder than the rich. New evidence (Flues and Thomas, 2015) for 21 OECD member countries shows that distributional effects of energy taxes differ by energy carrier. Taxes on transport fuels are generally not regressive, particularly when measured on an expenditure basis, largely because poorer households are less likely to use transport fuels. Taxes on heating fuels are found to be slightly regressive for various reasons. On the one hand, lower expenditure households may be particularly affected by taxes on heating fuels as they likely live in more poorly insulated dwellings. On the other hand, lower expenditure households are more likely to live in smaller dwellings with a smaller surface area to heat, and may conserve heating fuels by heating up to lower temperatures. Taxes on electricity are clearly regressive, which may be because it is difficult for poorer households to reduce electricity consumption.

Source: IPCC (2014), *Climate Change 2014: Mitigation of Climate Change*. Contribution of Working Group III to the Fifth Assessment Report of the Intergovernmental Panel on Climate Change, Cambridge University Press; Durand-Lasserve et al. (2015), "Modelling of distributional impacts of energy subsidy reforms: An illustration with Indonesia", *OECD Environment Working Papers*, No. 86, OECD Publishing, Paris; Flues, F. and A. Thomas (2015), "The distributional effects of energy taxes", *OECD Taxation Working Papers*, No. 23, OECD Publishing, Paris, *http://dx.doi.org/10.1787/5js1qwkqqrbv-en*.

around INR 200 billion (around USD 3 billion) between 2012-14. The retail price of diesel was increased periodically in small amounts (about INR 0.50 a month, or USD 0.008), eventually leading to the complete termination of diesel fuel subsidies in September 2014. To reduce the impact of this reform on the poor, the government increased the use of cash transfers, including subsidies for food, LPG and fertiliser. While large subsidies remain for kerosene and LPG (although much better targeted), the move represents an important step in the right direction (OECD, 2015d).

Mexico has eliminated its support for the consumption of gasoline and diesel fuel through its IEPS (*Impuesto Especial sobre Producción y Servicios por Enajenación de Gasolinas y Diesel*), a floating excise tax. The government uses international prices to set variable rates of IEPS for the country's two brands of gasoline, "Magna" and "Premium", and diesel fuel. When international oil prices are high, IEPS rates can turn negative, lowering domestic fuel prices and generating tax expenditures. Conversely, low international prices trigger an increase in IEPS rates, reducing tax expenditures, or, as is currently the case, resulting in a positive tax. In recent years, the Federal Mexican government has increased retail prices on a monthly basis. Together with lower international prices, these efforts have helped reduce total consumer support from MXN 244 billion (USD 18.5 billion) in 2012 to MXN 34 billion (USD 2.5 billion) in 2014 (OECD, 2015d).

Some other OECD member countries have also made progress on reforming fossil-fuel production and consumption subsidies on a smaller scale. The **Netherlands** phased out its excise-tax reduction for diesel fuel used for non-transport purposes (e.g. in farming activities or for heating) in January 2013. **Austria** and the **Slovak Republic** took similar steps in 2013 and 2011 respectively. **Canada** has reformed federal provisions relating to the treatment of certain capital expenses related to the oil sands, as well as for coal mining. **Germany** has reduced its budgetary transfers to hard-coal mines in North Rhine-Westphalia and plans to phase out these transfers entirely by 2018. **France** took steps in 2014 to gradually remove its excise tax exemption for natural gas consumed by households (OECD, 2015d).

References

ADB (2011), *Assessment and Implications of Rationalizing and Phasing Out Fossil-Fuel Subsidies*, Technical Assistance Report, *www.adb.org/projects/documents/assessment-and-implications-rationalizing-and-phasing-out-fossil-fuel-subsidies*.

APEC (2009), *A New Growth Paradigm for a Connected Asia-Pacific in the 21st Century*, Asia-Pacific Economic Co-operation, 14-15 November 2009, Singapore.

Arlinghaus, J. (2015), "Impacts of carbon prices on indicators of competitiveness: A review of empirical findings", *OECD Environment Working Papers*, No. 87, OECD Publishing, Paris, *http://dx.doi.org/10.1787/5js37p21grzq-en*.

CDC Climat (2013), Auction revenues in EU ETS Phase 3: a new public resource, *Climate Brief*, No. 25, *www.cdcclimat.com/IMG/pdf/13-01-24_climate_brief_no25_-_auction_revenues_in_eu_ets_phase_3.pdf*.

Durand-Lasserve, O., et al. (2015), "Modelling of distributional impacts of energy subsidy reforms: An illustration with Indonesia", *OECD Environment Working Papers*, No. 86, OECD Publishing, Paris, *http://dx.doi.org/10.1787/5js4k0scrqq5-en*.

Ecofys/World Bank (2014), *State and Trends of Carbon Pricing 2014*, World Bank Group, Washington, DC, *http://documents.worldbank.org/curated/en/2014/05/19572833/state-trends-carbon-pricing-2014*.

Ecologic Institute and eclareon (2014), *Assessment of Climate Change Policies in the Context of the European Semester, Country Report: Slovenia*, Ecologic Institute and eclareon, Berlin, *http://ec.europa.eu/clima/policies/g-gas/progress/docs/sl_2014_en.pdf*.

EDF/IETA (2014), *The World's Carbon Markets: A Case Study Guide to Emissions Trading: New Zealand*. Environmental Defence Fund and International Emissions Trading Association, March 2014, *www.ieta.org/assets/EDFCaseStudyMarch2014/new%20zealand%20case%20study%20march%202014.pdf*.

EEA (2015), *EU Emissions Trading System (ETS) Data Viewer* (database), European Environment Agency, *www.eea.europa.eu/data-and-maps/data/data-viewers/emissions-trading-viewer* (accessed 10 June 2015).

European Commission (2015), "Climate Action: Free Allocation Based on Benchmarks", webpage, *http://ec.europa.eu/clima/policies/ets/cap/allocation/index_en.htm* (accessed 10 June 2015).

European Commission (2014), *European Resource Efficiency Platform (EREP): Manifesto & Policy Recommendations*, Directorate-General for the Environment, Brussels, March 2014, *http://ec.europa.eu/environment/resource_efficiency/documents/erep_manifesto_and_policy_recommendations_31-03-2014.pdf*.

Flues, F. and B.J. Lutz (2015), "Competitiveness impacts of the German electricity tax", *OECD Environment Working Papers*, No. 88, OECD Publishing, Paris, *http://dx.doi.org/10.1787/5js0752mkzmv-en*.

Flues, F. and A. Thomas (2015), "The distributional effects of energy taxes", *OECD Taxation Working Papers*, No. 23, OECD Publishing, Paris, *http://dx.doi.org/10.1787/5js1qwkqqrbv-en*.

G-20 (2009), *Leaders' Statement*, The Pittsburgh Summit, 24-25 September 2009, *https://g20.org/wp-content/uploads/2014/12/Pittsburgh_Declaration_0.pdf*.

Galbraith, K. (2014), "Climate change concerns push Chile to forefront of carbon tax movement", *The New York Times*, 30 October 2014, *www.nytimes.com/2014/10/30/business/international/climate-change-concerns-push-chile-to-forefront-of-carbon-tax-movement.html?_r=0*.

Government of British Columbia (2014), "Carbon tax review" in *June Budget Update, 2013/14-2015/16*, *www.fin.gov.bc.ca/tbs/tp/climate/Carbon_Tax_Review_Topic_Box.pdf*.

Government of New Zealand (2015), *Friends of Fossil Fuel Subsidy Reform*, website, *www.mfat.govt.nz/fffsr/index.php* (accessed 20 June 2015).

Government of Ontario (2015), "Cap and trade system to limit greenhouse gas pollution in Ontario", 13 April 2015, *http://news.ontario.ca/opo/en/2015/04/cap-and-trade-system-to-limit-greenhouse-gas-pollution-in-ontario.html?utm_source=ondemand&utm_medium=email&utm_campaign=o*.

Harrison, K. (2013), "The political economy of British Columbia's carbon tax", *OECD Environment Working Papers*, No. 63, OECD Publishing, Paris, *http://dx.doi.org/10.1787/5k3z04gkkhkg-en*.

Henbest, S. (2015), "Fix the EU ETS, and carbon markets can be serious business", Bloomberg New Energy Finance.

IADB (2013), *Documento de Cooperación Técnica: Subsidios en el sector energía y políticas de mitigación al cambio climático en América Latina y el Caribe* [Technical Co-operation Document: Subsidies to the energy sector and climate-change mitigation policies in Latin America and the Caribbean], Inter-American Development Bank, *http://idbdocs.iadb.org/wsdocs/getdocument.aspx?docnum=38023936*.

ICAP (2015a), "China to cap emissions from six sectors, ETS to launch 2016", International Carbon Action Partnership, webpage, *https://icapcarbonaction.com/news/news-archive/268-china-to-cap-emissions-from-six-sectors-ets-to-launch-2016* (accessed 20 June 2015).

ICAP (2015b), International Carbon Action Partnership website, *https://icapcarbonaction.com/* (accessed 20 July 2015).

IEA (2014), *World Energy Outlook 2014*, IEA/OECD Publishing, Paris, *http://dx.doi.org/10.1787/weo-2014-en*.

IETA (2015), *The World's Carbon Markets: A Case Study Guide to Emissions Trading*, International Emissions Trading Association, Geneva, *www.ieta.org/worldscarbonmarkets* (accessed 20 April 2015).

IETA (2013), *A User Guide to Emissions Trading in China*, International Emissions Trading Association, Geneva, *www.ieta.org/a-user-guide-to-emissions-trading-in-china--september-2013*.

IPCC (2014), "Climate change 2014: Mitigation of climate change", in *Fifth Assessment Report of the Intergovernmental Panel on Climate Change*, Cambridge University Press, Cambridge, United Kingdom and New York, US.

Korea (Republic of) (2014), *First Biennial Update Report of the Republic of Korea to the United Nations Framework Convention on Climate Change*, December 2014.

Lanzi, E. et al. (2013), "Addressing competitiveness and carbon leakage impacts arising from multiple carbon markets: A modelling assessment", *OECD Environment Working Papers*, No. 58, OECD Publishing, Paris, *http://dx.doi.org/10.1787/5k40ggjj7z8v-en*.

Ministry of Sustainable Development and Energy of France (2014), Panorama énergies-climat – Edition 2014. La fiscalité de l'énergie [Panorama climate-energy - Edition 2014. Energy taxation], *www.developpement-durable.gouv.fr/IMG/pdf/32_-_La_fiscalit_de_l_nergie_-_Def.pdf*.

Moreira da Silva (2015), presentation at the OECD on Green Growth in Portugal, OECD Headquarters, Paris, 3 February 2015.

New Zealand (2015), Climate Change Information website, *https://www.climatechange.govt.nz/emissions-trading-scheme/obligations/* (accessed 20 April 2015).

OECD (2015a) *Database on Instruments used for Environmental Policy*, *www.2.oecd.org/ecoinst/queries/Default.aspx* (accessed 20 April 2015).

OECD (2015b), *Taxing Energy Use 2015: OECD and Selected Partner Economies*, OECD Publishing, Paris, *http://dx.doi.org/10.1787/9789264232334-en*.

OECD (2015c), *OECD Economic Surveys: China 2015*, OECD Publishing, Paris, *http://dx.doi.org/10.1787/eco_surveys-chn-2015-en*.

OECD (2015d), *OECD Companion to the Inventory of Support Measures for Fossil Fuels 2015*, OECD Publishing, Paris, *http://dx.doi.org/10.1787/9789264239616-en*.

OECD (2015e), *OECD Economic Surveys: Indonesia 2015*, OECD Publishing, Paris, *http://dx.doi.org/10.1787/eco_surveys-idn-2015-en*.

OECD (2015f), *OECD Economic Surveys: Japan 2015*, OECD Publishing, Paris, *http://dx.doi.org/10.1787/eco_surveys-jpn-2015-en*.

OECD (2015g), *OECD Economic Surveys: South Africa 2015*, OECD Publishing, Paris, *http://dx.doi.org/10.1787/eco_surveys-zaf-2015-en*.

OECD (2014a), *OECD Environmental Performance Reviews: Sweden 2014*, OECD Publishing, Paris, *http://dx.doi.org/10.1787/9789264213715-en*.

OECD (2014b), *OECD Economic Surveys: Australia 2014*, OECD Publishing, Paris, *http://dx.doi.org/10.1787/eco_surveys-aus-2014-en*.

OECD (2014c), *OECD Economic Surveys: India 2014*, OECD Publishing, Paris, *http://dx.doi.org/10.1787/eco_surveys-ind-2014-en*.

OECD (2013a), *Taxing Energy Use: A Graphical Analysis*, OECD Publishing, Paris, *http://dx.doi.org/10.1787/9789264183933-en*.

OECD (2013b), *Inventory of Estimated Budgetary Support and Tax Expenditures for Fossil Fuels 2013*, OECD Publishing, Paris, *http://dx.doi.org/10.1787/9789264187610-en*.

OECD (2011), "Interactions between emission trading systems and other overlapping policy instruments", General Distribution Document, Environment Directorate, OECD, Paris, *www.oecd.org/env/taxes*.

Republic of South Africa (2013), *Carbon Tax Policy Paper: Reducing Greenhouse Gas Emissions and Facilitating the Transition to a Green Economy*, National Treasury, *www.treasury.gov.za/public%20comments/Carbon%20Tax%20Policy%20Paper%202013.pdf*.

SEMARNAT (2014), "Carbon tax in Mexico", presentation at the World Bank partnership on market readiness (PMR), May 2014, *www.thepmr.org/system/files/documents/Carbon%20Tax%20in%20Mexico.pdf*.

Shen, Y. (2013), "Moving steadily or great leap forward? The emerging carbon market in China", *Deakin Law Review*, Vol. 18/2, Deakin University, Melbourne, pp. 233-270.

Swedish Tax Agency (2015), website, *www.skatteverket.se/servicelankar/otherlanguages/inenglish.4.12815e4f14a62bc048f4edc.html* (accessed 20 April 2015).

Swiss Federal Office for the Environment (2014), Emissions website, *www.bafu.admin.ch/emissionshandel/index.html?lang=en* (accessed 20 April 2015).

The Times of India, (2015), "Budget 2015: Power of five will light up energy sector", 1 March 2015, *http://timesofindia.indiatimes.com/budget-2015/union-budget-2015/Budget-2015-Union-Budget-2015/articleshow/46416518.cms*.

World Bank (n.d.), *Putting a Price on Carbon with a Tax*, *www.worldbank.org/content/dam/Worldbank/document/Climate/background-note_carbon-tax.pdf*.

World Bank (2015), "Carbon pricing initiatives valued at close to US$50 billion", 26 May 2015, *www.worldbank.org/en/news/press-release/2015/05/26/carbon-pricing-initiatives-valued-50-billion-dollars*.

Zhou, H. (2015), Allocation in China ETS, presentation at SinoCarbon innovation and investment, Mexico City, 27 January 2015, *http://climate.blue/wp-content/uploads/2015-01-27_DAY2_Presentation-Zhou_Allocation-in-the-Chinese-ETS.pdf*.

Climate Change Mitigation
Policies and Progress

Chapter 4

Policies in energy and other sectors

This chapter presents policies aimed at reducing greenhouse gas emissions in the power generation and transport sectors. The policy areas examined include renewable energy support policies, regulatory standards and innovation, and research and development. This chapter also looks at action to mitigate climate change in other sectors such as agriculture, land use, land-use change and forestry, industry and waste.

The statistical data for Israel are supplied by and under the responsibility of the relevant Israeli authorities. The use of such data by the OECD is without prejudice to the status of the Golan Heights, East Jerusalem and Israeli settlements in the West Bank under the terms of international law.

Policies to address emissions from power generation and transport

Carbon pricing instruments are an important instrument for reducing emissions in the power generation and transport sectors. At the same time, a growing number of other policy instruments are also regulating emissions from vehicles, power plants and buildings. This section provides an overview of renewable energy support policies, including feed-in tariffs, renewable portfolio standards and other fiscal policies. It also considers regulatory standards applied to vehicles, power plants and buildings, as well as home appliances.

Feed-in tariffs and premiums for renewable energy

Feed-in tariffs and premiums are used widely to support renewable electricity generation technologies. These mechanisms reduce risk for eligible generators by guaranteeing a pre-determined price for the electricity generated. Payments can be provided at a fixed level set independently of the wholesale electricity price ("feed-in tariffs"), or as a premium payment above the wholesale electricity price ("feed-in premiums"). Feed-in tariffs and premiums can be combined with price floors or caps to limit the risk of over- or under-compensation. They are also typically combined with guaranteed access to the grid for renewable generators.

Twenty-five OECD member countries and seven partner economies (the People's Republic of China [hereafter "China"], Costa Rica, India, Indonesia, Latvia, Lithuania and the Russian Federation) have feed-in tariffs and/or premiums in place (REN21, 2015). Their use has often significantly increased installed capacity of renewable energy in the countries concerned and has helped drive down the costs of renewable energy technologies (Criscuolo et al., 2014). Between 2008-12, for example, solar PV module prices fell by 80% and wind turbine prices fell by 29% due in part to the impact of feed-in tariffs and premiums (Liebreich, 2013). However, these policies can be costly; in some cases, the implicit carbon price of feed-in tariffs has reached hundreds of euros per tCO_2 (OECD, 2013a).

Getting the level of feed-in tariffs and premiums right has proven difficult. On the one hand, tariffs need to be predictable and high enough to reduce risk for investors and stimulate deployment of low-carbon technologies. On the other, they must avoid being over-generous and raising electricity prices too much for end consumers. In recent years, the financial implications of feed-in tariffs for government budgets have come under increasing scrutiny, leading to retroactive changes in levels of support in some cases (OECD, 2015a).

In **Spain**, the use of feed-in tariffs and premiums resulted in a rapid increase in the share of renewables in electricity generation. However, the costs of these policies were not fully passed onto consumers, resulting in a significant build-up of tariff debt owed to producers. Feed-in tariffs and other renewable energy incentives were changed to ensure sustainability of the system, although this has triggered some litigation with investors in renewable energy projects (OECD, 2014a). Similarly, the **Czech Republic** significantly reduced its feed-in tariffs and "green bonus" premium payments system for new renewable energy installations in 2014, although support remains available for some small-scale renewable energy systems. Other countries are also revising their feed-in tariffs (e.g. Greece, India, Italy, Japan, Lithuania and Portugal) (IEA, 2014c, 2013b).

Given its planned phase out of nuclear power generation by 2022, **Germany** is continuing to actively support renewable energy through a range of policy instruments, including feed-in tariffs. Following an extremely rapid expansion of renewable electricity capacity, particularly solar PV, the costs of this policy have risen significantly, reaching 0.8% of gross domestic product (GDP) in 2014. Since a surcharge on electricity bills funds the feed-in tariff prices, German households now face electricity prices considerably higher than in most neighbouring economies (OECD, 2014b). The feed-in tariff system was revised in 2014 and all tariffs decrease on a monthly basis at a rate that is linked to new installed capacity additions. Germany is now moving towards competitive bidding policies and the first auction of financial support for solar PV capacity took place in April 2015.

The **United Kingdom** launched an electricity market reform in 2013 partly to promote low-carbon energy sources through the Contracts for Difference scheme (OECD, 2015b). Contracts for Difference are a form of feed-in premium that provide revenue certainty for low-carbon electricity generators. Eligible low-carbon technologies include renewable energy sources, nuclear power and carbon capture and storage (CCS). The guaranteed prices of Contracts for Difference are determined via auctions. At the first auction in February 2015, guaranteed prices ranged from GBP 50 per MWh for certain solar PV projects to GBP 120 per MWh for certain offshore wind projects (DECC, 2015).

Some countries have shifted from feed-in tariffs to feed-in premiums. Feed-in premiums have now been introduced in an increasing number of European countries, including the Czech Republic, Denmark (for onshore wind), Estonia, Germany, the Netherlands, Slovenia and Spain. Some of these countries offer renewable energy generators a choice between a feed-in tariff and a feed-in premium.

In 2014, the European Union approved new guidelines on state aid for environmental protection and energy 2014-20 (2014/C 200/01). These stipulate that aid for electricity from renewable energy sources should be granted through a competitive bidding process for allocating public support as of January 2017. As a result, competitive bidding processes and feed-in premiums will progressively replace feed-in tariffs (except for small installations) (European Commission, 2014a).

Renewable portfolio standards and tradable certificates

Renewable portfolio standards (RPSs) are mandates for electric utilities to source a fixed share or quantity of electricity from renewable sources. They are often accompanied by tradable renewable energy certificate programmes to enhance the cost effectiveness of compliance. Thirteen OECD member countries and five partner economies have implemented RPSs or quotas. They are also used widely at the sub-national level – 54 states and provinces in the United States, Canada and India had RPSs in place as of 2014 (REN21, 2015).

The largest renewable portfolio standard system is in the **United States:** 29 states require electric utilities to supply a fixed share of electricity from renewables, or install an equivalent amount of renewable electricity generation capacity (DSIRE, 2015). The levels for these standards vary widely – from 33% by 2020 in California to 2% by 2021 in South Carolina. There are further design differences in aspects such as eligibility, treatment of existing projects, treatment of energy imports and exports, compliance methods and degree of regulatory oversight (Barbose, 2012). A cost-benefit analysis estimated the incremental cost of these state policies (i.e. compared to scenarios without the standards) at less than 1% of average retail electricity rates in 2010-12 (Heeter et al., 2014). Despite these low costs, many renewable portfolio standards are facing legal challenges or review.

Korea replaced its feed-in tariff by a renewable portfolio standard in 2012. The new system ensures that large power generation companies supply an increasing share of electricity from renewables. This share, which started from 2% in 2012, will increase up to 10% in 2024 (Korea, 2014; Kemco, n.d). **Mexico** created a tradable Clean Energy Certificate system in 2014, partly modelled after California's renewable portfolio standards. In other countries, RPSs have recently been reformed or phased out. For example, the **United Kingdom** Renewables Obligation will be phased out in 2017; the decision to establish an RPS for solar power in the state of Tamil Nadu in **India** was overturned in 2013 (REN21, 2014).

Some countries have introduced green certificate trading systems for renewable energy, such as Australia, Austria, Belgium, Czech Republic, Denmark, Finland, France, Ireland, Italy, Japan, Netherlands, Norway, Poland, Slovak Republic, Slovenia, South Korea, Spain, Sweden, United Kingdom and the United States. India also has a Renewable Energy Certificate (REC) system (REN21, 2014).

Australia is pursuing its Renewable Energy Target (RET) of producing 23.5% of electricity supply using renewable resources by 2020 through two programmes: one for large-scale business such as energy suppliers (the Large-scale Renewable Energy Target, LRET) and one for owners of small-scale systems (the Small-scale Renewable Energy Scheme, SMRES). Energy retailers are required to purchase renewable energy certificates from renewable energy generators which can then be traded. In 2015, the LRET was reduced from 41 000 GWh to 33 000 GWh in 2020 with interim and post-2020 targets adjusted accordingly as part of the Renewable Energy (Electricity) Amendment Bill 2015 (Clean Energy Regulator, 2015).

In **Belgium**, energy suppliers are obliged to purchase certificates issued to "green" producers to promote the production of electricity from renewable energy sources and combined heat and power, both at federal and regional levels. Due to the advantageous financial support, public interest in installing photovoltaic panels turned out to be higher than expected. As a result, the initial distribution rules of the certificates were modified to avoid market imbalances (Government of Belgium, 2013).

Sweden introduced an electricity certificates system to promote electricity from renewable energy sources in 2003. Under the system, which partly overlaps with the EU ETS, eligible electricity generators receive one electricity certificate for every megawatt-hour of renewable electricity produced; these certificates are then sold to electricity retailers and end users. The share of certificates that has to be purchased by users is gradually increasing up to 2020. In 2012, this market mechanism was extended to **Norway** with a joint target of increasing renewable energy production (Ministry of Environment of Sweden, 2014).

Poland is replacing its green certificate trading scheme by auctions to increase the use of renewables in electricity generation, which remains low. Since 2005, the green certificate trading system had incentivised many state-owned coal-fired power plants to co-fire with biomass, but the development of other technologies has been limited (OECD, 2014c). In February 2015, the Polish Parliament adopted new legislation on renewable energy sources that promotes the use of auctions as of 2016.

Fiscal and financial incentives

Around 30 of the countries studied (including 8 partner economies) have reduced certain tax rates to promote renewable energy; more than 32 countries have introduced capital subsidies or rebates; and 23 countries have set investment or production tax credits (REN21, 2015). A federal Investment Tax Credit in the **United States** reduces federal income tax for persons or companies investing in residential and commercial solar PV systems. This tax credit is set to decrease from 30% to 10% in 2017 (SEIA, n.d). In 2013, **Luxembourg** began offering grants for projects to improve energy savings and use renewable energy sources in the private building sector (IEA/IRENA, 2015).

In 2014, **Colombia** enacted Energy Law 1715 to promote the development of renewable energy sources in the national energy system. It foresees action plans to encourage energy use of agricultural and forest biomass, solid waste unsuitable for reuse and recycling, and wind resource generation projects, as well as to explore the potential of geothermal and solar energy. It sets financial instruments such as the Non-Conventional Energy and Efficient Energy Management Fund to support these projects and fiscal incentives such as tax deductions (LSE-GRI, 2015).

Since the 1990s, **Norway** has supported the sale of electric vehicles through different fiscal and financial measures such as a zero registration tax, a zero toll policy and a zero VAT. The vehicle registration tax shifted towards a system that rewards vehicles with low CO_2 emissions and penalises vehicles with high emissions. The average CO_2 emission from new cars was reduced from 177 g/km in 2006 to 123 g/km 2013 (Government of Norway, 2014). The registration tax rate is defined according to the weight of the car and the engine size, on top of which CO_2 and NO_x components were introduced in 2007 and 2012. Tax benefits for electric vehicles were to be unchanged until 2017, unless the number of zero emission cars exceeded 50 000. Following the introduction of new electric vehicles in the market, Norway reached its 50 000th car in April 2015 and is consequently reviewing its fiscal incentives for electric vehicles (Reuters, 2015). As Norway relies on renewable energy to generate most of its electricity, shifting to electric cars contributes to reducing GHG emissions.

In March 2015, **Costa Rica** announced the Acquisition Programme for Efficient Vehicles (PAVE) to promote increased fuel economy of cars. PAVE subsidises buyers of new cars in a certain category (MINAE, 2015). Since its electricity sector is largely decarbonised due to the high share of hydro and geothermal power, Costa Rica will need to significantly reduce transport emissions to meet its carbon neutrality goal.

Regulatory standards

Regulatory standards are used widely to address GHG emissions from the transport, power and buildings sectors. This section outlines observed trends in the use of regulatory standards such as fuel economy standards for vehicles, emission standards for power plants and energy efficiency standards for buildings. Unlike direct carbon pricing policies, regulatory standards put an implicit price on carbon. Ensuring that climate policy mixes are cost effective is important (OECD, 2014d; Box 4.1).

Since it can take decades to turn over the vehicle fleet and building stock, there can be a time lag between introducing regulatory standards and reducing GHG emissions. Thus, in many cases, regulatory standards are likely to have a small impact on GHG emissions levels in the short term, and a larger impact in future. Further, rebound effects (i.e. increased energy use resulting from increased efficiency) can undo part of the reductions achieved through regulatory standards.

Box 4.1. **Maximising the cost effectiveness of climate policy mixes**

Given the pressures on government budgets, it is unlikely that emissions will be significantly reduced unless governments pursue cost-effective climate policy mixes. To that end, the best instruments are explicit carbon pricing mechanisms, such as carbon taxes and emissions trading systems (ETSs). Maximising coverage can improve the cost effectiveness of these policy instruments, while ensuring coherence with other policies and the same implicit or explicit carbon price for all emitters.

OECD (2013a) found that regulatory approaches are less cost effective than direct carbon pricing instruments such as taxes and trading systems. The prices effectively placed on carbon provide insights into the cost effectiveness of carbon abatement policies. The Productivity Commission of Australia and the OECD have shown that, in practice, carbon abatement prices vary according to the sector, the country and the instrument used (OECD, 2013a). Policy instruments can nonetheless be implemented for other purposes than reducing GHG emissions.

There is great variation between the effective carbon abatement prices of different policy instruments. In the electricity generation sector, most countries were found to have a carbon price of at least EUR 25 per tCO_2 thanks to different incentives. The OECD estimated it can cost up to EUR 800 to abate 1 tonne of carbon according to the policy instrument. With respect to abating carbon, capital subsidies and feed-in tariffs systems were found to be the least cost-effective instruments, whereas trading systems and broad-based taxes were the most cost effective. Carbon prices found in the pulp and paper and cement sectors were relatively modest. The household sector is subject to various abatement incentives, many of which rank above EUR 100 per tonne of carbon abated; again, taxes on energy products are the most cost-effective instruments.

In the road transport sector, the price of carbon abatement also varies according to instrument. Regulatory standards to support biofuels in the transport sector were found to be much more costly than fuel taxes on an abatement cost basis. For instance, some policies resulted in a carbon price exceeding EUR 1 000 per tonne of carbon abated. The average effective carbon price for fuel mandates in the countries studied was 441 EUR per tCO_2e abated, compared with 55 EUR per tCO_2e abated for fuel taxes. The carbon prices implicit in regulations such as performance standards typically exceed the explicit carbon price of market-based instruments.

Source: OECD (2013a), *Effective Carbon Prices*, OECD Publishing, Paris, *http://dx.doi.org/10.1787/9789264196964-en*.

Fuel economy standards and biofuel mandates

In the road transport sector, fuel economy standards are used to improve the fuel efficiency of new vehicles. These standards now cover 70% of the global passenger light-duty vehicle market. New standards are expected to save between USD 40-190 billion in fuel costs by 2020 (IEA, 2014a).

The **United States** has adopted strict national fuel economy standards for light-duty cars and trucks. The US Environmental Protection Agency (EPA) and Department of Transportation jointly issued Corporate Average Fuel Economy (CAFE) standards for new light-duty vehicles in 2012. The CAFE standards are expected to result in an average vehicle performance of 5.2 litres (L) per 100 km for model year 2017-25 passenger cars and light trucks (US Department of State, 2014). For comparison, the average new car sold in the United States in 2012 achieved 7.8 L per 100 km in fuel economy. The United States has also finalised fuel efficiency standards for commercial trucks, vans and buses, and has proposed a second phase of fuel standards for medium- and heavy-duty vehicles for model year 2018 and beyond (EPA, 2015c). In keeping with US standards, **Canada** underlines in its INDC that it has established more stringent GHG emission standards for passenger automobiles and light trucks. A Canadian 2025 model year passenger vehicle should emit half as many emissions as a 2008 model.

The **European Union** has introduced stringent fuel economy standards for new light-duty vehicles. From 159 gCO_2 per km in 2007, the standard for new passenger cars will drop to 130 gCO_2 per km by 2015 (equivalent to around 5.6 L per 100 km of petrol or 4.9 L per 100 km of diesel) and to 95 gCO_2 per km by 2021 (4.1 L per 100 km of petrol or 3.6 L per 100 km of diesel). From 203 gCO_2 per km in 2007, the standard for new vans will drop to 175 gCO_2 per km by 2017 and to 147 gCO_2 per km by 2020 (European Commission, 2014b). Standards for the post-2020 period are to be developed by the end of 2015. Complementing the new standards, the EU will introduce CO_2 labelling legislation to increase the awareness of consumers. The EU has yet to set fuel economy standards for heavy-duty vehicles such as trucks and buses. Switzerland has regulations aligned with the EU's emission standards.

Japan's Top Runner programme includes a set of efficiency standards for products, including vehicles, heaters and various electrical appliances. The policy, in place since 1999, aims to trigger a "race to the top" among manufacturers (Kodaka, n.d.). The scope, which has been expanded over time, covers 31 products (METI, 2014; Energy Conservation Center Japan, n.d.). Producers have between 3-10 years to meet standards equal to the performance of the most efficient product in the class available on the market (taking into account expected improvements). In the case of passenger cars, the Top Runner programme exceeded expectations – a 22.8% increase in fuel economy relative to 1995 levels was achieved by 2005, five years ahead of schedule (Kimura, 2012).

China and **Korea** have upgraded their fuel economy standards. China's national fuel efficiency standards are being developed in stages. Stage 1 sets a fuel use limit of 8.9-10.1 L per 100 km (depending on the weight of the vehicle) to be met by 2006; Stage 2 sets a limit of 8.1-9.2 L per 100 km to be met by 2009; and Stage 3 sets a limit of 6.5-7.3 L per 100 km to be met by 2015. A further standard for passenger cars of 5 L per 100 km has been proposed for 2020 (UNEP, 2015). Korea also tightened its fuel economy standards for automobiles. From 2012-15, cars and multi-purpose passenger vehicles should not emit more than 140 g of CO_2/km; from 2016-20, permitted emissions drop to 97 g of CO_2/km (Korea, 2014).

Standards are also being established for the share of renewable fuels in total fuel use. Blending mandates are used to increase the use of biofuels in transport. Biofuels, used as an alternative to fossil fuels in transport, are liquid or gaseous transport fuels (biodiesel and bioethanol) made from biomass. The **EU** 2009 Renewable Energy Directive (European Commission, 2009a) set "a mandatory 10% minimum target to be achieved by all Member States for the share of biofuels in transport petrol and diesel consumption by 2020, to be introduced in a cost-effective way". Fuel suppliers must also reduce the greenhouse gas intensity of the EU fuel mix by 6% by 2020 compared to 2010 as defined by the 2009 Fuel Quality Directive (European Commission, 2009b). This is to be obtained partly through the use of alternative fuels such as biofuels.

Some countries are also developing advanced biofuel technologies, commonly referred to as second- or third-generation, or "advanced biofuels". This category includes biofuels based on lignocellulosic biomass, such as cellulosic-ethanol, biomass-to-liquids

(BtL)-diesel and bio-synthetic gas (bio-SG). The category also includes novel technologies that are mainly in the research, development and demonstration (RD&D) and pilot stage, such as algae-based biofuels and the conversion of sugar into diesel-type biofuels using biological or chemical catalysts.

Other examples of policy support to biofuels include the following:

- **Brazil** increased its minimum ethanol content mandate from 25% to 27.5% and has a bioethanol blend mandate of 25% (REN21, 2015). Since 2014, Brazil has had a biodiesel blend mandate of 7%.
- **California** introduced a low carbon fuel standard that requires providers to reduce the carbon intensity of transportation fuels by 10% by 2020. Fuel providers can meet their obligations in part through use of alternative fuels, including second-generation biofuels (California EPA, 2015).
- Nine provinces in **China** have a 10% ethanol mandate for transport. Overall, the country seeks to move to a 10% biofuels mandate by 2020 (Shiyan et al., 2012).
- One of **Colombia's** mitigation goals is to increase the share of biofuels in the fuel mix by 20% by 2020. It intends to stimulate the growth of biofuel production such as ethanol and biodiesel without endangering natural forests or food security (UNFCCC, 2013). A biofuels mandate was introduced in 2012.
- **Costa Rica** removed its biofuels mandate, but retained mandates for ethanol and biodiesel blends remain (REN21, 2015).
- At the end of 2014, **Italy** became the first country to require advanced biofuels in cars and trucks. From 2018, all fuel suppliers in the country will have to include 0.6% advanced biofuels in petrol and diesel (McGrath, 2014).This rate will increase to 0.8% in 2020 and to 1% in 2022.

Emission standards for power plants

Performance standards for power plants are reducing the carbon intensity of the electricity mix. This intensity depends on the share of low-carbon electricity generation technologies such as renewables and nuclear power, as well as the carbon intensity of fossil fuels used in fossil-fired plants. The carbon intensity of electricity generation in the OECD as a whole declined by 14% between 1990 and 2012 (from 498 gCO_2 per kWh to 427 gCO_2 per kWh) (Figure 4.1).

The **United States** and **Canada** are using performance standards to reduce emissions from power plants. The US EPA is developing emission standards for new and existing power plants as part of President Obama's 2013 Climate Action Plan under the Clean Air Act. The proposals, which include emission standards for new and existing power plants, are differentiated by technology type (Box 4.2). In Canada, the federal government has established an emission standard for coal-fired power plants. The new standard of 420 gCO_2 per kWh will accelerate phase out of coal-fired electricity generation units and in effect ban the construction of traditional coal-fired plants without CCS. Further, the Canadian province of **Ontario** made a commitment in 2002 to phase out the use of coal for electricity generation by 2015. This objective was achieved in 2014 and follow-up legislation has since been passed to ban coal from ever being used for electricity generation in Ontario after 2014 (Government of Ontario, 2014).

Figure 4.1. **Carbon intensity of electricity generation**

Grammes of CO_2 per kilowatt hour, 2012

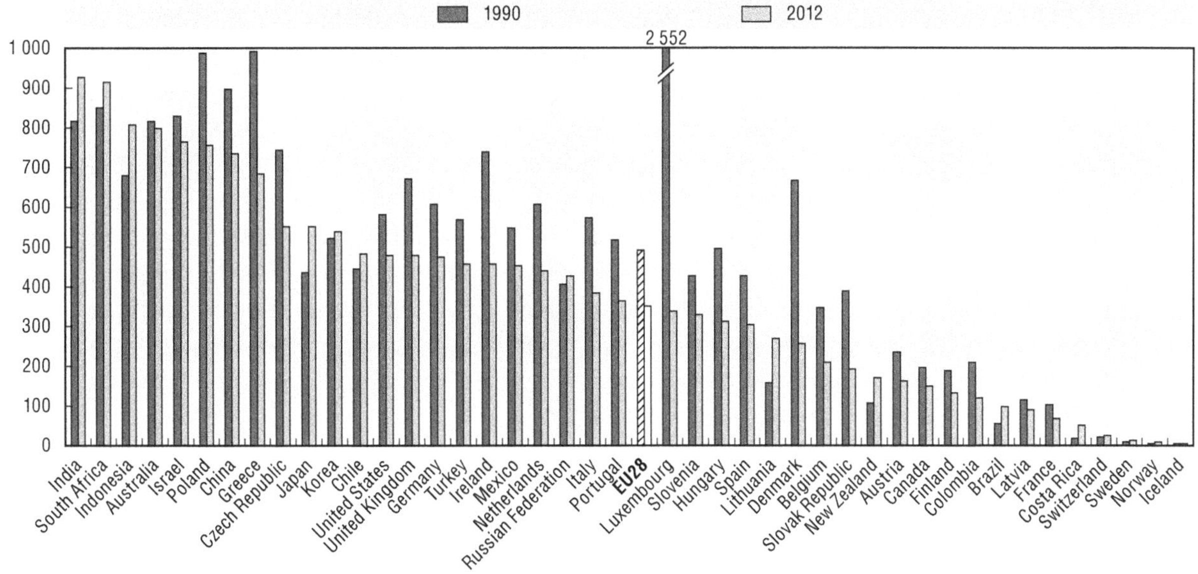

Source: IEA (2015b), "Emissions per kWh of electricity and heat output", *IEA CO_2 Emissions from Fuel Combustion Statistics* (database), IEA/OECD, (accessed 21 April 2015).

StatLink http://dx.doi.org/10.1787/888933272500

Box 4.2. **The US EPA's emission standards for new and existing power plants**

As part of a national effort to address CO_2 emissions from power plants under the Clean Air Act, the US finalised in August 2015 a Clean Power Plan to cut carbon pollution from existing power plants as well as final carbon pollution standards for new, modified and reconstructed power plants. The 2015 Clean Power Plan sets an overall target to reduce CO_2 emissions from the power sector by 32% below 2005 levels by 2030. To reach this target, the plan includes a federal plan, state-specific CO_2 goals and a model rule to assist states in implementing the Clean Power Plan.

For new power plants, emission standards have been set to reflect the best system of emission reduction (BSER) for each class of plant. The emission standard has been set at 454 gCO_2/kWh (1 000 lb CO_2/MWh) for new and reconstructed gas-fired power plants, 635 gCO_2/kWh (1 400 lb CO_2/MWh) for new coal-fired power plants and 816 to 907 gCO_2/kWh (1 800 to 2 000 lb CO_2/MWh) for reconstructed coal-fired power plants.

Sources: EPA (2015a), *Carbon Pollution Emission Guidelines for Existing Stationary Sources: Electric Utility Generating Units*; EPA (2015b), *Standards of Performance for Greenhouse Gas Emissions from New, Modified, and Reconstructed Stationary Sources: Electric Utility Generating Units*.

Energy efficiency standards for buildings and appliances

Energy efficiency measures are used to reduce emissions from buildings in the residential, industrial and commercial sectors. A growing number of countries have established energy efficiency standards for new buildings and home appliances. However, the slow turnover rate of housing stock means that policies targeting energy efficiency in existing buildings (known as "retrofitting") are also needed. Numerous countries are implementing efficiency standards for home appliances to promote energy efficiency.

Information and labelling measures are also used to influence behaviour and encourage more energy-efficient practices.

The **European Union**, the **United States** and **Canada** have established strict building codes for new buildings. The EU Energy Performance Building Directive 2010/31/EU requires EU Member States to apply minimum energy performance requirements for new and existing buildings. The US Department of Energy's Building Energy Codes Programme (BECP) develops new building codes and provides technical support for meeting them. The building codes developed under the BECP are expected to improve the energy efficiency of new buildings by 50% (US Department of State, 2014). Canada's National Energy Code for Buildings (NECB), established in 2011, outlines minimum energy efficiency levels for building envelopes; systems and equipment for heating; ventilating and air conditioning; service water heating; lighting; and the provision of electrical power systems and motors. The code, which applies to all new buildings, is 25% more stringent than the previous building code (Government of Canada, 2014).

Numerous countries have established standards for home appliances. White goods such as refrigerators and washing machines can be highly energy-intensive, but some products are more efficient than others. Endorsement labels contribute to a "race to the top" in best products, while energy efficiency standards for appliances remove certain products from the market. For instance, under the 2002 Minimum Energy Performance Standards (MEPS) of **New Zealand**, new products entering the market must meet or exceed minimum standards before they can be sold in the country. This regulation applies to appliances such as dishwashers, air conditioners and refrigerators, and regular updates increase the number of products covered.

Countries have widely used labelling and other programmes to improve information available to consumers on how to save energy. Although the impact of such measures may be difficult to quantify, they can contribute to behaviour change. Labelling programmes have been most commonly used for home electric appliances and thermal insulation of buildings. The **United States** and the **European Union**, for example, have introduced regulations that mandate the energy performance labelling of home appliances. The US Energy Star programme is a voluntary energy labelling scheme first launched by the US EPA in 1992 under the Clean Air Act. It applies to energy efficient products in 65 categories; commercial buildings (more than 20 000 buildings in the United States are top performers, using 35-40% less energy than average buildings); 24 industrial sectors (by 2012, 120 plants were awarded an Energy Star label); and new homes (US Department of State, 2014). This programme has been extended to the European Union, which has also adopted an Energy Labelling Directive (2010/30/EU); this defines labelling requirements for specific product groups such as air conditioners, boilers, televisions and washing machines. Certificates are also used to provide information on consumption in buildings and appliances.

Domestic offset schemes

Switzerland requires importers of fossil fuels and operators of fossil thermal power plants to compensate domestically for their CO_2 emissions. For power plants, at least half of the compensation must come from projects within the country. For fuel importers, the share of emissions to be compensated increases annually from 2% in 2014-15 to at least 10% by 2020. The resulting compensation costs may not exceed CHF 0.05 per L. To be eligible for tradable attestations, domestic compensation projects must be voluntary measures that go beyond legal requirements. Domestic compensation projects in the following categories are eligible: energy efficiency, renewable energy, mobility management, wood products and biofuels with high quality standards.

In parallel with the planned carbon tax in **South Africa**, the National Treasury is also considering a carbon offset scheme. This offset scheme will introduce the possibility to reduce carbon tax liability up to 5-10% depending on the sector by investing in external GHG reduction projects in South Africa; it will allow carbon credits to be issued, in specific projects or activities that reduce, avoid or sequester emissions. The mechanism would provide flexibility to heavy users and lower the cost to the economy in terms of growth (OECD, 2015d).

Costa Rica launched a Voluntary Domestic Carbon Market (MDVCCR) in 2013, as part of its efforts to become carbon neutral. The MDVCCR establishes guidelines for the generation, issuance and exchange of carbon credits (known as Costa Rican Compensation Units, UCCs) from activities originating in Costa Rica. Eligible activities include reforestation and energy efficiency projects.

Innovation and RD&D

Innovation plays an important role in broadening the range of low-carbon technology options available to governments and the private sector over time. In the power sector, these options include the next generation of renewable electricity generation technologies, nuclear power and CCS, as well as energy storage technologies and smart grid technologies. In the transport sector, new low-carbon vehicles are being developed for road transport, rail, waterborne transport and aviation, including vehicles that run on electricity, hydrogen fuel cells, compressed or liquefied natural gas, and sustainably-produced biofuels. The extent to which these technologies reduce emissions depends on the feedstocks and conversion processes (IPCC, 2014). In the buildings sector, advanced building materials and energy-efficient home appliances are being developed and existing technologies improved. The industrial sector needs to switch to lower-carbon and alternative fuels for production; make more efficient use of materials; and deploy best available technologies to the greatest extent possible in all sub-sectors, as well as CCS for deeper decarbonisation.

Many of the countries studied have developed RD&D strategies that outline research priorities. In the United States, the 2011 Blueprint for a Secure Energy Future aims to position the country as a global leader in innovation and the development and manufacturing of clean energy technologies. The strategy includes research funding and grants. The European Commission, European Council and European Parliament regularly prepare the Multi-annual Framework Programme for Research and Innovation, which sets out general and sector-specific objectives, the amount of spending per sector and spending rules. The latest research and innovation framework programme – Horizon 2020 – runs from 2014-20. The EU's goal is to increase total RD&D spending (both public and private) from less than 2% of GDP in 2010 to 3% of GDP in 2020 (IEA, 2014d, 2014e). Australia introduced a set of Science and Research Priorities, including for energy and environmental change, to ensure that appropriate levels of public funding are allocated to research. Each priority also has related "practical challenges" such as aiming for low emission energy production, reliable efficient and cost-effective storage technologies, and options for responding and adapting to the impacts of environmental change on biological systems, urban and rural communities and industry (Government of Australia, 2015).

Public RD&D plays a key role in accelerating technological improvements and driving down the costs of low-carbon technologies. In the absence of public intervention, the private sector would likely under-fund RD&D due to the fact that positive spill-over effects are often not fully captured by investors. Private industries can also struggle to protect their intellectual property and overcome the "valley of death" between the early and late stages of technology development. The IEA conducts comprehensive reviews of developments in energy technologies (IEA, 2015a).

Targeted policies are needed on both the supply and demand sides, and at all stages of the research, development, demonstration and deployment chain. Support is needed for basic research, applied research, demonstration projects, scaled-up deployment and commercial diffusion into the marketplace. Different policies may be appropriate at different stages of the RD&D chain. Direct grants and public research programmes are often needed at the early stages. Tax incentives, feed-in tariffs and regulatory standards can accelerate deployment and commercialisation of more mature technologies at later stages. Government departments and research laboratories can also play an important role in convening industry members to address RD&D.

Statistics on public energy-related RD&D spending in OECD member countries are not available for all years or all technologies. Furthermore, RD&D statistics for the ten partner economies studied are unavailable. Most of the available statistics focus on public RD&D; by definition, this fails to capture private sector efforts. The following information and charts present trends in public energy-related RD&D spending in OECD member countries.

In 2012, 22 OECD member countries spent around USD 13 billion on public energy-related RD&D (IEA, 2015c). The share of public energy-related RD&D in total public RD&D expenditure has been declining since the 1980s (IEA, 2014c). Public energy-related RD&D spending peaked in absolute terms after the global financial crisis, as several governments directed economic stimulus funds towards energy-related RD&D projects, but remains low as a share of GDP. Among the OECD member countries for which statistics are available, total energy-related RD&D accounted for between 0.001% (Portugal in 2013) and 0.167% (Luxembourg in 2012) of national GDP, with the simple average around 0.045% (Figure 4.2).

Figure 4.2. **Total public energy RD&D spending**

Percentage share of nominal GDP

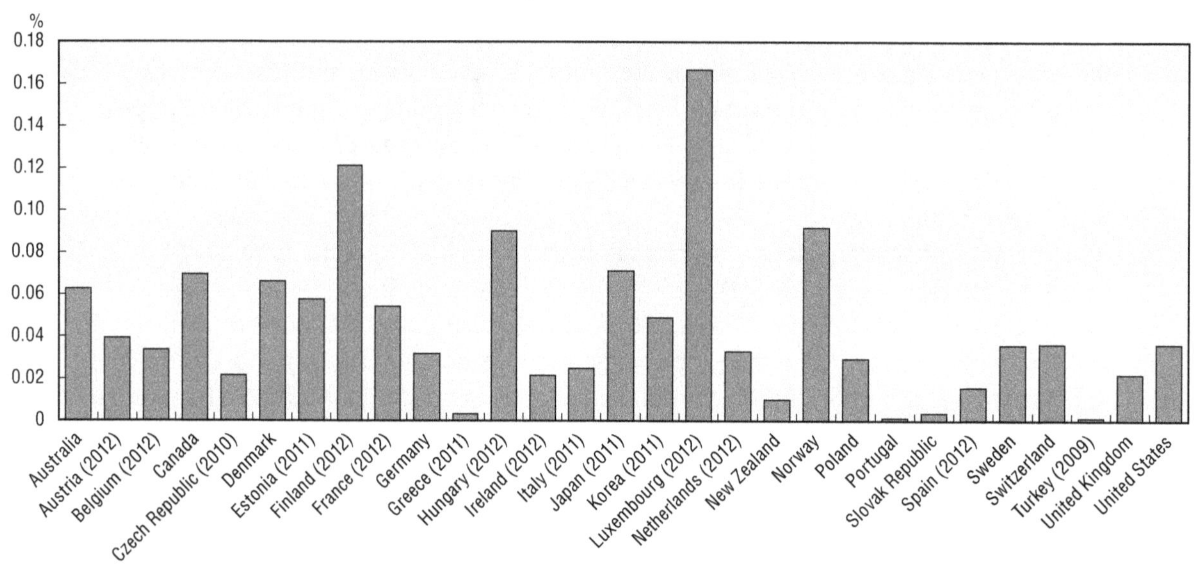

Note: Data for 2013 except where indicated otherwise.

Source: IEA (2015c), "RD&D Budget", *IEA Energy Technology RD&D Statistics* (database), *http://dx.doi.org/10.1787/data-00488-en* (accessed 21 July 2015); IEA (2015d), "RD&D Indicators", *IEA Energy Technology RD&D Statistics* (database), *http://dx.doi.org/10.1787/data-00489-en* (accessed 21 July 2015).

StatLink http://dx.doi.org/10.1787/888933272510

RD&D for low-carbon technologies

OECD member countries have focused their energy-related RD&D spending on various technologies (Figure 4.3). In the OECD member countries for which statistics were available in 2012, the largest allocations of public energy-related RD&D spending were to renewable energy sources (24%), energy efficiency (20%) and nuclear energy (16%). Further, in the countries for which data are available, the shares of renewable energy and energy efficiency in public energy-related RD&D spending are rising, while the share of nuclear power has declined since its peak in the 1980s.

Figure 4.3. **Public energy RD&D spending by flow**

Percentage of energy-related RD&D expenditure, 2013

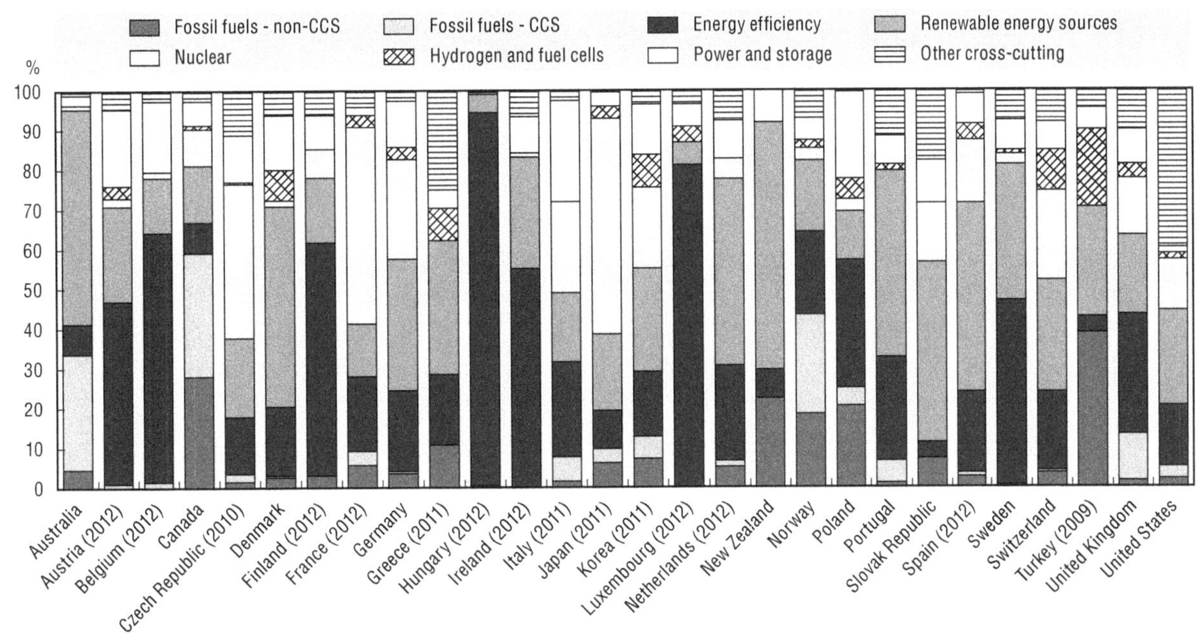

Note: Based on total public RD&D, including government, public agencies and state-owned enterprises. Statistics for 2013 until indicated otherwise.

Source: IEA (2015c), "RD&D Budget", *IEA Energy Technology RD&D Statistics* (database), *http://dx.doi.org/10.1787/data-00488-en* (accessed 21 July 2015).

StatLink *http://dx.doi.org/10.1787/888933272520*

For renewable energy, the main areas of public RD&D spending are biofuels and solar energy. The countries with the highest share of renewable energy technologies in total public energy-related RD&D spending are New Zealand (62% in 2013, mainly biofuels and geothermal), Australia (54% in 2013, mainly solar), Denmark (50% in 2013, mainly wind energy), the Netherlands (47% in 2013, mainly biofuels and solar) and Spain (47% in 2012, mainly solar and biofuels).

For energy efficiency, the main areas of public RD&D spending are transport (particularly road vehicles and battery technologies), industrial techniques and processes, and residential and commercial buildings, appliances and equipment. The countries with the highest levels of spending on energy efficiency as a share of total public energy-related RD&D spending are Hungary (94% in 2012, mainly industry), Luxembourg (81% in 2012, mainly road vehicles and advanced combustion engines), Belgium (62% in 2012, mainly industry), Finland (59% in 2012, mainly industry and residential and commercial buildings, appliances and equipment) and Ireland (55% in 2012, mainly residential and commercial buildings, appliances and equipment).

Public RD&D spending on nuclear power as a share of total public energy-related RD&D is highest in Japan (54% in 2011), France (49% in 2012), the Czech Republic (39% in 2010) and Germany (25% in 2013). The main areas of public RD&D spending on nuclear energy are plant safety and integrity, the fuel cycle and nuclear breeder reactors. In addition to nuclear fission technologies, governments are also supporting RD&D of nuclear fusion technologies for the second half of the century.

Carbon capture and storage accounts for over half of public RD&D spending on fossil fuels in OECD member countries for which data are available, and approximately USD 22 billion of direct financial support was provided to large-scale CCS projects between 2008-12 (IEA, 2014c). Twenty-two large-scale CCS projects are currently in operation or under construction in various industries worldwide. Of these, the first large-scale CCS project in the power sector commenced operation in Canada in 2014 (Boundary Dam power station in the province of Saskatchewan); two further large-scale CCS projects in the power sector are due to begin operation in the United States in 2015 and 2016 (Global CCS Institute, 2014). Australia is one of the countries providing significant support for CCS. In 2009 the Australian Government established the Global CCS Institute. The Australian government also has a range of support programmes like the National Low Emissions Coal Initiative (NLECI) and the CCS Flagships programmes that foster the development of low-emissions technologies and support CCS demonstrations projects such as the Otway storage demonstration project and the Gorgon CCS project in Western Australia.

Smart grid technologies include advanced information, communications and control systems that make more efficient use of energy on both the demand and the supply side, improve resiliency to disturbances, enable greater integration of renewables, and raise awareness of energy consumption among consumers through smart meters. Global smart-grid technology investments reached USD 45 billion in 2013, up from USD 33 billion in 2012 (IEA, 2014c). While most activity has revolved around deployment of smart meters at the consumer end, there has been a recent surge in investment and innovative activity in automation and control of distribution grids (IEA, 2015a).

Private sector involvement in RD&D for low-carbon technologies

The private sector remains an important source of funding for energy-related RD&D, although statistics relating to private energy-related RD&D expenditures are incomplete (Haščič and Migotto, 2015). Global energy-related RD&D spending by industry was estimated to reach around USD 22 billion in 2014 (Batelle, 2013). In **Finland**, businesses accounts for around 70% of total energy-related RD&D expenditure. Finland has set up several programmes to support energy-related RD&D from the private sector, such as the Groove-Growth from Renewables 2010-14 programme, to enhance the capabilities of small and medium-sized companies working with renewable energy (IEA, 2013a).

In **Germany**, the private sector spent about EUR 300 million on energy-related research in 2009 (IEA, 2013b). The government provides a maximum of 50% funding for projects run by commercial enterprises to support application-oriented research. In **Estonia**, public funding for energy-related RD&D as a share of GDP (0.058% in 2011) is relatively high compared to the average for OECD member countries (around 0.045%), but private funding is still limited. As a result, nearly all basic research is conducted in the public sector, while the private sector focuses on product development and innovation (IEA, 2013c). Through the **Korean** Green Energy Strategy Roadmap, the private sector is expected to contribute 50% of RD&D costs and be the principal instigator of large-scale projects (IEA, 2012).

Data on patent applications can reflect inventive performance. However, not all technologies or processes are the subject of applications, and not all enterprises wish to disclose technological advances through patent applications (OECD, 2015c). Among OECD member countries in 2010, energy efficiency accounted on average for 0.7% total patents under the Patent Cooperation Treaty. Hungary had the highest share of energy efficiency patents (close to 7% of total patents), followed by the Netherlands (2.5%) and Austria (1.6%). Among partner economies, energy efficiency patents represented 0.4% of China's patents, followed by South Africa (0.3%) and India (0.1%) (OECD, 2015c).

Policies to address emissions from non-energy sectors

In most countries, the energy sector – including energy industries and transport – accounts for the largest share of GHG emissions. However, other sectors such as agriculture, land use and forestry, industry and waste are important sources or sinks of GHG emissions in some countries, particularly in partner economies. Since 1990, GHG emissions from industry, agriculture and waste have declined in the OECD as a whole.

Mitigation abatement costs vary between sectors. Ultimately, emissions will need to be reduced in all key emitting sectors to achieve deep cuts in global GHG emissions. Market mechanisms should provide incentives to mitigate emissions in sectors with low abatement costs. At the same time, non-market policy instruments (such as incentives for innovation) could help reduce abatement costs in other sectors over longer periods.

Agriculture

Agricultural activities are an important source of GHG emissions in many countries, particularly methane (CH_4) and nitrous oxide (N_2O) emissions from livestock, soil and nutrient management (Smith et al., 2014). Agriculture accounts for 8% of the OECD's total GHG emissions, although the share of agriculture in national GHG emissions varies from 2% in Japan to 46% in New Zealand. Given that agriculture is highly sensitive to climate change, particularly changes in temperature and rainfall patterns, many countries are developing both adaptation and mitigation policies in the sector. Implementing mitigation policies in this sector has proven challenging partly due to the current lack of cost-effective agricultural mitigation technologies in many regions.

The emission intensity of agricultural output is improving in most, but not all, OECD member countries. The contribution of climate policies to this achievement, if any, remains unclear. Many countries and areas have achieved an absolute decoupling of agricultural economic growth and GHG emissions. Emissions from agriculture in OECD member countries have declined slightly over the past two decades, from 1 380 to 1 332 $MtCO_2e$ in 1990-2012, while total agricultural output and productivity increased slightly in 1990-2010 (OECD, 2014e). Although decoupling trends show that agriculture has become less emissions-intensive, absolute emissions from agriculture continue to rise in some countries studied, including the United States, Canada, New Zealand and Turkey.

Cost-effective practices to reduce emissions from the agriculture sector exist in some regions. These include improved efficiency of fertiliser use, cattle breeding and energy efficiency in mobile machinery, as well as education and information measures (Macleod et al., 2015). Emissions can also be reduced with improved farming efficiency, use of alternative energy sources, and enhanced soil and vegetation CO_2 sinks (Wreford et al., 2010).

Some countries have introduced policies to reduce GHG emissions from agriculture and are moving in the right direction. In most cases, however, countries have taken limited action to tackle emissions from this sector. Some countries are developing agricultural climate change programmes to reduce GHG emissions, while making agriculture more resilient to climate change. Mitigation policies have generally focused on changes to livestock management, manure management and optimising use of nitrogen fertilisers.

There is growing interest in the concept of "climate-smart" agriculture, which aims to combine mitigation, adaptation and productivity objectives. A Global Alliance on Climate Smart Agriculture was launched at the UN Climate Summit in 2014. The Alliance seeks to adjust agricultural, forestry and fisheries practices, food systems and social policies so they better take account of climate change and the efficient use of natural resources (FAO, 2015). Eleven of the 44 countries studied in this report are members of the Alliance: Costa Rica, France, Japan, Mexico, the Netherlands, Norway, South Africa, Spain, Switzerland, the United Kingdom and the United States. The Alliance also includes the UN Secretary General office, the Food and Agriculture Organization, the World Bank, and other inter-governmental organisations, research institutions and non-governmental organisations. Members of the Alliance have agreed to enhance co-operation to achieve the following three aspirational outcomes: (i) sustainable and equitable increases in agricultural productivity and incomes; (ii) greater resilience of food systems and farming livelihoods; and (iii) reduction and/or removal of GHG emissions (FAO, 2014).

The **EU's** GHG emissions from agriculture have fallen since 1990, but reductions are not expected to continue without more effort. Reductions were due to better productivity and farm practices, as well as fewer cattle. The main potential to continue reducing emissions lies in improved livestock and manure management. According to one study, improved nitrogen fertiliser management had the largest impact of any measure on the European Union's N_2O emissions (European Commission, 2009c). The EU's Common Agricultural Policy (CAP) provides incentives to modernise farms, for instance, by improving energy efficiency. The policy provides direct farm payments according to environmental conditions. The new CAP for the period 2014-20 will further enhance the existing policy framework for sustainable management of natural resources, contributing to both climate change mitigation and enhancing the resilience of farming to the threats posed by climate change and variability. Member States need to include "agri-environment-climate" measures in their Rural Development Programmes under which direct farm payments are made according to environmental conditions.

Some governments are establishing partnerships with non-state actors to pursue mitigation measures on the demand side, as well as the supply side. In **Ireland**, for example, the Irish Food Board introduced the Origin Green voluntary programme in 2012 to promote sustainability in agriculture. Participants are required to submit five-year plans, including emission targets. The voluntary programme is based on carbon footprint monitoring that was introduced for beef farms (OECD, 2013b). The **Netherlands** has a programme called the Green Deals to encourage the private sector to reduce emissions in agriculture. The national government, the Dutch Dairy Organisation and the Dutch Agricultural and Horticultural Organisation have set a target to reduce carbon emissions of dairy chains to zero by 2020 (OECD, 2013b).

Carbon pricing instruments rarely cover emissions from agriculture. **New Zealand** originally planned to include agriculture in its ETS, but delayed implementation several times. Since 2012, however, agricultural facilities such as meat processors, dairy processors, fertiliser manufacturers and importers, and live animal exporters are required to report on GHG emissions; they do not yet have to surrender permits. In **Australia**, the Carbon Farming Initiative was established in 2011 to enable farmers and land managers to generate tradable credits for GHG mitigation. Following a change of government, the initiative was integrated into Australia's new Emissions Reduction Fund in December 2014 (Clean Energy Regulator, 2014).

Although many INDCs cover gases from agriculture, most countries have not specified which actions they will undertake to reduce emissions from this sector. China states it will make efforts to achieve zero growth of fertiliser and pesticide use by 2020 to promote low-carbon development in agriculture. China intends to control methane emissions from rice fields and nitrous oxide emissions from farmland and to build a recyclable agriculture system, promoting comprehensive use of straw, reuse of agricultural and forestry wastes and comprehensive use of animal waste (UNFCCC, 2015a). Other countries such as Iceland, which relies on renewable energy for its electricity and heating, emphasise that it will have to focus on other sectors for mitigation options, including agriculture and fisheries (UNFCCC, 2015a).

Land use, land-use change and forestry

Forestry and other land use can range from being a significant source of GHG emissions to a significant carbon sink in different countries. For example, CO_2 removals from **Latvia's** large forested areas are greater than its GHG emissions from other sectors, making the country a net GHG sink. In other countries such as **Brazil** and **Indonesia**, deforestation can be a major source of GHG emissions, although these countries have reduced deforestation rates in recent years. While the exact causes of deforestation vary from place to place, a key underlying driver is increasing demand for food, fuel and forest products such as timber. For this reason, policies to tackle deforestation are closely inter-linked with other policy areas, particularly concerning agriculture.

Reducing emissions from deforestation and forest degradation (REDD) is an approach in the UNFCCC negotiations for results-based payments to developing countries for achieving reductions in deforestation and forest conservation relative to a baseline level (Karousakis and Corfee-Morlot, 2007). The term "REDD+" also takes into account the conservation of forest carbon stocks, sustainable forest management and enhancement of forest carbon stocks. The REDD approach puts a price on the various ecosystem services provided by forests, including their capacity to mitigate climate change. **Brazil** became the first developing country to submit its proposed forest baseline level to the UNFCCC in 2014. Since then, **Colombia** and **Mexico** have also submitted baseline proposals (UNFCCC, 2015b).

The deforestation rate has slowed in **Brazil**, although an area of forest the size of Belgium is still destroyed every five to six years (OECD, 2013c). As a result, Brazil's total GHG emissions, including LULUCF, have substantially decreased since 2004. The decrease in deforestation rates was largely due to the Action Plan for the Prevention and Control of Deforestation in the Legal Amazon (PPCDAm). This federal programme, launched in 2004, aims to reduce the annual rate of deforestation by 80% by 2020. An action plan is also in place to reduce deforestation in the Cerrado biome.

Indonesia has the world's third-largest forest coverage (944 320 km^2, or 52% of the land area). However, its forest cover has decreased by around 20% since 1990 (Figure 4.4). Fibre production, logging and palm oil were found to be the primary causes of deforestation in

Figure 4.4. **Trends in extent of forest**

Percentage change 1990-2010

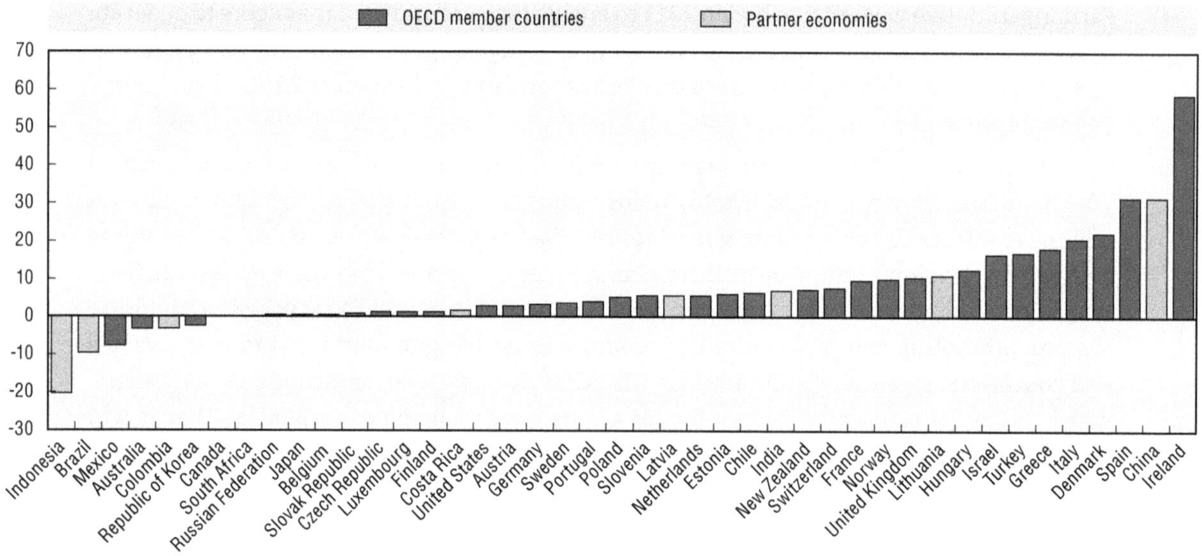

1. Iceland not shown on this chart.

Source: FAO (2010), "Trends in extent of forest 1990-2010" in *Global Forest Resources Assessment 2010*, *www.fao.org/forestry/fra/fra2010/en/*.

StatLink http://dx.doi.org/10.1787/888933272532

Indonesia (International Sustainability Unit, 2015). Further, an estimated 89% of the nation's 128 million ha are under no regulation or permit, making them difficult to protect (OECD, 2015e). Indonesia's policies to address deforestation focus on strengthening regulatory measures against irresponsible behaviours (such as illegal logging and forest burning, which has been banned since 1995), preventing forest fires, promoting forest-related RD&D and tree replanting in urban areas (Government of Indonesia, 2012).

In most OECD member countries, the main focus of policies for enhancing mitigation in the LULUCF sector has been on sustainable forest management practices and on replanting of previously deforested areas. Sustainable forest management refers to the use of forests and forest land in a way and at a rate that maintains their biodiversity, productivity, regeneration capacity and vitality. It ensures the potential remains to fulfil ecological, economic and social functions at local, national and global levels. Forests and other wooded land cover more than 40% of the EU's land area; close to a quarter of this area is protected under Natura 2000. The **European Union** adopted a new Forest Strategy in 2013 that includes several initiatives to support, implement and assess sustainable forest management. The EU strategy also includes support for REDD+ and measures to tackle illegal logging in the world's forests (European Commission, 2013).

Forestry was the first sector to be covered by the **New Zealand** ETS in 2008. Under the ETS, owners of land forested prior to 1990 are eligible to surrender emission allowances if deforestation has occurred. Owners can purchase emission allowances, either buying international units or paying a fixed price of NZD 25 tCO_2e to meet their deforestation obligations. Since January 2013, some forest landowners can offset deforestation on their land by planting an equivalent area of new forest elsewhere in New Zealand (Government of New Zealand, 2015).

Industry

GHG emissions from industrial activity stem from the use of energy, as well as from industrial processes. Under the UNFCCC, the industrial processes sector refers to direct GHG emissions from activities such as the production of cement, iron, steel, aluminium and manufacturing industries. Indirect emissions associated with the energy inputs to industrial processes are included under the energy sector for the purpose of GHG inventories under the UNFCCC. Emissions have decreased since 1990 in all OECD member countries, but still account on average for 8% of national emissions. This decline has been offset by growing industrial emissions in partner economies, mostly in Asia (Bernstein et al., 2014). Cement production is the most important source of non-energy industrial process emissions. However, other activities also contribute such as the production of iron and steel, aluminium, chemicals, textiles and leather, chemicals, and pulp and paper.

Most emissions trading systems and energy and carbon taxes to date have covered a significant share of industrial sources of emissions; an exception is the United States Regional Greenhouse Gas Initiative, which only covers the power sector. Other policy instruments to address emissions from the industrial sector include information programmes, tradable energy efficiency certificate schemes, and regulations and standards. The majority of policies to date aim at improving industrial energy efficiency or reducing process emissions. Regulations address emissions of CO_2, as well as other GHGs such as fluorinated gas. These so-called F-gases, such as hydrofluorocarbons (HFCs), perfluorocarbons (PFCs) and sulphur hexafluoride (SF_6), are used in a range of industrial applications.

Information programmes are widely used to promote energy efficiency in industry. The most common programmes are data collection systems, auditing, and measurement and reporting requirements (Bernstein et al., 2014). Auditing initiatives can apply to residential, commercial and public buildings, as well as to industrial facilities.

Auditing and reporting requirements were part of **China's** Top-1 000 Energy-Consuming Enterprises Programme. The programme set energy consumption targets for 1 000 of China's largest industrial companies, including steel mills, chemical factories and coal mines. Together, these companies accounted for around one-third of China's total energy use. The programme surpassed its energy-saving target of 100 million tonnes of coal equivalent (Mtce), with reported savings of over 150 Mtce. It was subsequently scaled-up in the 12th Five-Year Plan (2011-15) to cover 10 000 enterprises, accounting for two-thirds of China's total energy consumption. It also includes financial incentives to promote energy-saving retrofits (Institute for Industrial Productivity, n.d).

In addition to its Top-10 000 programme, China has introduced regulations to promote transformation and upgrading of traditional industries, as well as the closure of small and inefficient production facilities. In 2013, for example, China mandated the closure of facilities that generate 6.2 million tonnes of iron, 8.8 million tonnes of steel, 270 000 tonnes of aluminium and 106 million tonnes of cement, while more than 2 000 small coal mines are to be closed by 2015 (NDRC, 2014). These regulations reflect China's "new normal" economic paradigm, which adjusts the economic structure by tackling overcapacity in polluting heavy industries and shifting from energy-intensive industries towards service industries.

In **India**, the National Mission for Enhanced Energy Efficiency includes multiple programmes aimed at industrial GHG emissions, including the Market Transformation for Energy Efficiency, the Energy Efficiency Financing Platform and the Perform, Achieve and Trade Mechanism for Energy Efficiency (PAT). Under the PAT, energy consumption targets are allocated to energy-intensive industries. Installations that exceed their energy savings targets are rewarded with tradable certificates. As of 2012, the PAT applied to nearly 500 industries, including aluminium, cement, iron and steel producers. The mechanism is expected to save 26.21 $MtCO_2e$ in its first three years (2012-15) (Government of India, 2012; Pahuja et al., 2014).

In the **United States**, regulations under the Clean Air Act mandate industrial plants to use "best available control technologies" to control GHG emissions as part of the Prevention of Significant Deterioration programme (EPA, 2015d). In **Japan**, the Act on the Rational Use of Energy (amended in 2010) introduced mandatory energy efficiency targets that were based on benchmarks. Japan has also established a voluntary initiative called the Keidanren Commitment to Low Carbon Society. The 2010 **EU** Industrial Emissions Directive aims to minimise the pollution from various industrial sources through permits. It promotes the use of "best available techniques" to reduce emissions (European Commission, 2010). Further, new EU legislation to address emissions of fluorinated GHGs from industrial processes was passed in 2014. Regulation (EU) No 517/2014 limits the total amount of certain fluorinated gases that can be sold in the EU ETS and bans the use of fluorinated gases in many types of new equipment such as refrigerators or air conditioning systems (European Commission, 2014b).

Waste

In the OECD, the share of waste in total national GHG emissions ranges from 0.4% in Luxembourg to 12% in Portugal. Between 1990-2012, 20 OECD member countries reduced the level of emissions from waste, while 13 others increased it. GHG emissions from waste depend on the amount of waste and the way it is disposed of, i.e. landfilling, recycling or incineration. In national GHG inventories under the UNFCCC, the waste sector includes land disposal of solid waste, wastewater handling and nitrous oxide from human sewage. The major GHG emissions are landfill CH_4, followed by wastewater CH_4 and N_2O (Bogner et al., 2007). Mitigation policies in the sector that can help reduce emissions include preventing waste, promoting waste recycling, and recovering or capturing and combustion of landfill gas. Sustainable materials management can also abate emissions significantly in the production, consumption and end-of-life of materials (OECD, 2012).

In the **European Union**, 42% of waste per capita is recycled, 36% is sent to landfill and 22% is incinerated (European Commission, 2014b). EU policies and measures on waste comprise several key directives. In particular, the overarching Waste Framework Directive 2008/98/EC establishes a legal framework for waste management. This directive introduces the polluter pays principle and the principle of extended producer responsibility. It also promotes the waste management hierarchy: (i) prevention; (ii) recycling; (iii) energy recovery from waste; and (iv) disposal. The Landfill Directive 1999/31/EC set targets to reduce the landfilling of biodegradable municipal waste, and mandates the collection of landfill gas from all landfills receiving biodegradable waste.

Fiscal incentives such as taxes on solid waste disposal are widely implemented. Governments are also revising tax rates to internalise the cost of environmental damage. For example, **Estonia** is planning to revise taxes on oil-shale related activities that result in GHG-emitting waste (OECD, 2015f). **Ireland** has one of the highest levels of waste generation per capita in the OECD coupled with a high reliance on landfills to dispose waste (OECD, 2013d). The Irish government raised the landfill levy from EUR 65 to EUR 75 per tonne in 2013 to reduce disposal of waste to landfill (Government of Ireland, 2014). In **Latvia**, however, increases in tax rates have not resulted in significant diversion of waste from landfill to recycling (OECD, 2015g). **Hungary** ratified an Act on Waste in 2012. This act, which puts a focus on "life-cycle thinking", introduces a landfill fee that rises annually from EUR 10.7 per tonne to EUR 42.8 per tonne in 2013-16. While climate change mitigation was not the primary objective of this policy, the legislation is expected to result in GHG emissions reduction from the waste sector (Government of Hungary, 2013).

Landfill waste regulations are widely used in OECD member countries. **Sweden** introduced a ban on landfill disposal in 2002, which has contributed to reduce waste emissions (Ministry of the Environment of Sweden, 2014). In the **United States**, large solid waste landfills are mandated to capture and combust their landfill gas emissions (US Department of State, 2014). Twenty-one of the 44 countries studied are members of the Global Methane Initiative. This initiative, launched in 2004, now covers around 70% of global CH_4 emissions. It aims to enhance international collaboration on methane abatement, recovery and use in five areas: agriculture, coal mines, municipal solid waste, oil and gas systems, and wastewater (Global Methane Initiative, 2015).

References

Barbose, G. (2012), "Renewables portfolio standards in the United States: A status update", presentation at the 2012 National Summit on RPS, Washington, DC, 3 December 2012, *www.cesa.org/assets/2012-Files/RPS/RPS-SummitDec2012Barbose.pdf.*

Batelle (2013), *2014 Global R&D Funding Forecast*, December 2013, *www.battelle.org/docs/tpp/2014_global_rd_funding_forecast.pdf?sfvrsn=4.*

Bernstein, L. et al. (2014), "Industry" in *Climate Change 2007: Mitigation. Contribution of Working Group III to the Fourth Assessment Report of the Intergovernmental Panel on Climate Change*, Metz, B. et al. (eds) Cambridge University Press, Cambridge, United Kingdom and New York, US, *https://www.ipcc.ch/pdf/assessment-report/ar4/wg3/ar4-wg3-chapter7.pdf.*

Bogner, J. et al. (2007), "Waste management", in *Climate Change 2007: Mitigation. Contribution of Working Group III to the Fourth Assessment Report of the Intergovernmental Panel on Climate Change*, Metz, B. et al. (eds). Cambridge University Press, Cambridge, United Kingdom and New York, US.

California EPA (2015), "Low Carbon Fuel Standard Programme", California Environmental Protection Agency, webpage, *www.arb.ca.gov/fuels/lcfs/lcfs.htm* (accessed 20 June 2015).

Clean Energy Regulator (2015), "About the Renewable Energy Target," webpage, *www.cleanenergyregulator.gov.au/RET/About-the-Renewable-Energy-Target* (accessed 24 August 2015).

Clean Energy Regulator (2014), "Transitions to the Emissions Reduction Fund", webpage, *www.cleanenergyregulator.gov.au/ERF/Transition-to-the-Emissions-Reduction-Fund* (accessed 20 June 2015).

Criscuolo, C. et al. (2014), "Renewable energy policies and cross-border investment: Evidence from mergers and acquisitions in solar and wind energy", *OECD Science, Technology and Industry Working Papers*, No. 2014/03, OECD Publishing, Paris.

DECC (2015) "Contracts for difference allocation round one outcome, United Kingdom Department of Energy and Climate Change", *https://www.gov.uk/government/uploads/system/uploads/attachment_data/file/407059/Contracts_for_Difference_-_Auction_Results_-_Official_Statistics.pdf.*

DSIRE (2015), "Renewable portfolio standard policies", Database of State Incentives for Renewables and Efficiency, NC Clean Technology Center, NC State University, Raleigh, US, *http://ncsolarcen-prod.s3.amazonaws.com/wp-content/uploads/2014/11/Renewable-Portfolio-Standards.pdf.*

Energy Conservation Center Japan (n.d.) "Top runner programme" (in Japanese), *https://www.eccj.or.jp/machinery/toprunner/toprunner.pdf* (accessed 08 July 2015).

EPA (2015a), *Carbon Pollution Emission Guidelines for Existing Stationary Sources: Electric Utility Generating Units. www.2.epa.gov/sites/production/files/2015-08/documents/cpp-final-rule.pdf* (accessed 24 August 2015).

EPA (2015b), *Standards of Performance for Greenhouse Gas Emissions from New, Modified, and Reconstructed Stationary Sources: Electric Utility Generating Units. www.epa.gov/airquality/cpp/cps-final-rule.pdf* (accessed 24 August 2015).

EPA (2015c), *EPA and NHTSA Propose Standards to Reduce Greenhouse Gas Emissions and Improve Fuel Efficiency of Medium- and Heavy-Duty Vehicles for Model Year 2018 and Beyond*, *www.epa.gov/OMS/climate/documents/420f15901.pdf* (accessed 25 August 2015).

EPA (2015d), *Air Enforcement*, *http://www2.epa.gov/enforcement/air-enforcement* (accessed 31 August 2015).

European Commission (2014a), "Guidelines on state aid for environmental protection and energy 2014-2020", *Communication from the Commission*, 28 June 2014, *http://eur-lex.europa.eu/legal-content/EN/TXT/?uri=CELEX:52014XC0628(01)*.

European Commission (2014b), *Sixth National Communication and First Biennial Report from the European Union under the UN Framework Convention on Climate Change*, *http://unfccc.int/files/national_reports/annex_i_natcom_/application/pdf/eu_nc6.pdf*.

European Commission (2013), "A new EU forest strategy: For forests and the forest-based sector", *Communication from the Commission — Guidelines to the European Parliament, the Council, the European Economic and Social Committee and the Committee of the Regions*, COM (2013) 659 final, SWD (2013) 343 final, *http://ec.europa.eu/agriculture/forest/strategy/staff-working-doc_en.pdf*.

European Commission (2010), "Directive 2010/75/EU of the European Parliament and of the Council of 24 November 2010 on industrial emissions (integrated pollution prevention and control).

European Commission (2009a), "Directive 2009/28/EC of the European Parliament and of the Council of 23 April 2009 on the promotion of the use of energy from renewable sources and amending and subsequently repealing Directives 2001/77/EC and 2003/30/EC", *http://eur-lex.europa.eu/legal-content/EN/ALL/?uri=CELEX%3A32009L0028*.

European Commission (2009b), "Directive 2009/30/EC of the European Parliament and of the Council of 23 April 2009 amending Directive 98/70/EC as regards the specification of petrol, diesel and gas-oil and introducing a mechanism to monitor and reduce greenhouse gas emissions and amending Council Directive 1999/32/EC as regards the specification of fuel used by inland waterway vessels and repealing Directive 93/12/EEC", *http://eur-lex.europa.eu/legal-content/EN/TXT/?uri=CELEX%3A32009L0030*.

European Commission (2009c), *Commission Staff Working Document: The Role of European Agriculture in Climate Change Mitigation*, 23 July 2009, 1093 final, *http://ec.europa.eu/agriculture/climate-change/pdf/sec2009_1093_en.pdf*.

FAO (2015), Climate Smart Agriculture website, *www.fao.org/climate-smart-agriculture/85725/en/* (accessed 08 July 2015).

FAO (2014), "Global alliance for climate-smart agriculture (GACSA) framework document", Food and Agriculture Organization, Rome, Version 1, 1 September 2014, *www.fao.org/climate-smart-agriculture/41760-02b7c16db1b86fcb1e55efe8fa93ffdc5.pdf*.

FAO (2010), "Trends in extent of forest 1990-2010" in *Global Forest Resources Assessment 2010*, *www.fao.org/forestry/fra/fra2010/en/*.

Global CCS Institute (2014), *The Global Status of CCS: 2014*, Global CCS Institute, Melbourne, Australia, *http://decarboni.se/sites/default/files/publications/180923/global-status-ccs-2014.pdf*.

Global Methane Initiative (2015), website, *https://www.globalmethane.org/* (accessed 20 June 2015).

Government of Australia (2015), *Science and Research Priorities*, May 2015. *www.science.gov.au/scienceGov/ScienceAndResearchPriorities/Documents/15-49912%20Fact%20sheet%20for%20with%20National%20Science%20and%20Research%20Priorities_4.pdf* (accessed 24 August 2015).

Government of Belgium (2013), *Belgium's Sixth National Communication to the United Nations Framework Convention on Climate Change*, *https://unfccc.int/files/national_reports/annex_i_natcom/submitted_natcom/application/pdf/bel_nc6_rev_eng.pdf*.

Government of Canada (2014), *Canada's Sixth National Communication to the United Nations Framework Convention on Climate Change*, *http://unfccc.int/national_reports/annex_i_natcom_/items/1095.php*.

Government of Hungary (2013), *Hungary's Sixth National Communication on Climate Change*, *http://unfccc.int/files/national_reports/annex_i_natcom/submitted_natcom/application/pdf/nc6-final_hun[1].pdf*.

Government of India (2012), *India's Second National Communication to the United Nations Framework Convention on Climate Change*, *http://unfccc.int/resource/docs/natc/indnc2.pdf*.

Government of Indonesia (2012), *Indonesia's First National Communication to the United Nations Framework Convention on Climate Change*, *http://unfccc.int/resource/docs/natc/indonc1.pdf*.

Government of Ireland (2014), *Ireland's Sixth National Communication and First Biennial Report to the United Nations Framework Convention on Climate Change.*

Government of New Zealand (2015), "Climate Change Information", webpage, *https://www.climatechange.govt.nz/.*

Government of Norway (2014), *Norway's Sixth National Communication to the United Nations Framework Convention on Climate Change.*

Government of Ontario (2014), Ontario's Climate Change Update 2014, Ministry of the Environment and Climate Change, Toronto, Canada, *https://dr6j45jk9xcmk.cloudfront.net/documents/3618/climate-change-report-2014.pdf.*

Heeter, J. et al. (2014), "A survey of state-level cost and benefit estimates of renewable portfolio standards", May 2014, National Renewable Energy Laboratory, Golden, US and Lawrence Berkeley National Laboratory, Berkeley, US, *www.nrel.gov/docs/fy14osti/61042.pdf.*

IEA (2015a), *Energy Technology Perspectives 2015*, IEA/OECD Publishing, Paris, *www.iea.org/etp/.*

IEA (2015b), "Emissions per kWh of electricity and heat output", *IEA CO_2 Emissions from Fuel Combustion Statistics* (database), *http://dx.doi.org/10.1787/data-00432-en* (accessed 21 July 2015).

IEA (2015c), "RD&D Budget", *IEA Energy Technology RD&D Statistics* (database), *http://dx.doi.org/10.1787/data-00488-en* (accessed 21 July 2015).

IEA (2015d), "RD&D Indicators", *IEA Energy Technology RD&D Statistics* (database), *http://dx.doi.org/10.1787/data-00489-en* (accessed 21 July 2015).

IEA/IRENA (2015), *Global Renewable Energy Policies and Measures* (database), *www.iea.org/policiesandmeasures/* (accessed 08 July 2015).

IEA (2014a), *Energy Efficiency Market Report 2014*, *www.iea.org/bookshop/463-Energy_Efficiency_Market_Report_2014.*

IEA (2014c), *Tracking Clean Energy Progress 2014*, IEA/OECD Publishing, Paris, *www.iea.org/etp/tracking/.*

IEA (2014d), *Energy Policies of IEA Countries: United States 2014*, IEA/OECD Publishing, Paris, *http://dx.doi.org/10.1787/9789264211469-en.*

IEA (2014e), *Energy Policies of IEA Countries: European Union 2014*, IEA/OECD Publishing, Paris, *http://dx.doi.org/10.1787/9789264190832-en.*

IEA (2013a), *Energy Policies of IEA Countries: Finland 2013*, IEA/OECD Publishing, Paris, *http://dx.doi.org/10.1787/9789264190788-en.*

IEA (2013b), *Tracking Clean Energy Progress 2013*, IEA/OECD Publishing, Paris, *www.iea.org/publications/tcep_web.pdf.*

IEA (2013c), *Energy Policies Beyond IEA Countries: Estonia 2013, IEA, Paris, http://dx.doi.org/10.1787/9789264190801-en.*

IEA (2012), *Energy Policies of IEA Countries: Republic of Korea 2012*, IEA/OECD Publishing, Paris, *http://dx.doi.org/10.1787/9789264171497-en.*

Institute for Industrial Productivity (2015), *Industrial Efficiency Policy* (database), *http://iepd.iipnetwork.org/* (accessed 20 April 2015).

International Sustainability Unit (2015), *Tropical Forests. A Review*, April 2015, The Prince of Wales's International Sustainability Unit (ISU).

IPCC (2014), "Climate change 2014: Mitigation of climate change", in *Fifth Assessment Report of the Intergovernmental Panel on Climate Change*, Cambridge University Press, Cambridge, UK and New York, US.

Haščič, I. and M. Migotto (2015), "Measuring environmental innovation using patent data", *Environment Working Paper*, No. 89, *www.oecd.org/officialdocuments/publicdisplaydocumentpdf/?cote=ENV/WKP(2015)10&docLanguage=En.*

Karousakis, K. and J. Corfee-Morlot (2007), *Financing Mechanisms to Reduce Emissions from Deforestation: Issues in Design and Implementation*, OECD Publishing, Paris, *www.oecd.org/env/cc/39725582.pdf.*

Kemco (n.d.), Korean Energy Management Corporation website, *www.kemco.or.kr/new_eng/pg02/pg02040705.asp* (accessed 20 June 2015).

Kimura, O. (2012), "The role of standards: The Japanese top runner program for end-use efficiency", *Historical Case Studies of Energy Technology Innovation*, 12, *www.iiasa.ac.at/web/home/research/researchPrograms/TransitionstoNewTechnologies/12_Kimura_Japan_TopRunner_WEB.pdf*.

Kodaka (n.d), "Japan's top runner program: The race for the top", Ministry of Economic Trade and Industry, *www.eceee.org/events/eceee_events/product_efficiency_08/programme_presentations/Kodaka_TopRunnerProgram.pdf* (accessed 20 April 2015).

Korea (Republic of) (2014), *First Biennial Update Report of the Republic of Korea to the United Nations Framework Convention on Climate Change*, December 2014, *http://unfccc.int/resource/docs/natc/rkorbur1.pdf*.

Liebreich, M. (2013), "Keynote address", presentation at the Bloomberg financial summit, New York, 23 April 2013, *http://about.bnef.com/presentations/bnef-summit-2013-keynote-presentation-michael-liebreich-bnef-chief-executive/*.

LSE-GRI (2015), "Law 1715, regulating the integration and promotion of non-conventional renewable energy to the national energy system", *The Global Climate Legislation Database*, London School of Economics – Grantham Research Institute on Climate Change and the Environment, *www.lse.ac.uk/GranthamInstitute/law/law-1715-regulating-the-integration-and-promotion-of-non-conventional-renewable-energy-to-the-national-energy-system/* (accessed 10 July 2015).

MacLeod, M., et al. (2015), "Cost-Effectiveness of Greenhouse Gas Mitigation Measures for Agriculture: A Literature Review", *OECD Food, Agriculture and Fisheries Papers*, No. 89, OECD Publishing, Paris, *http://dx.doi.org/10.1787/5jrvvkq900vj-en*.

McGrath, M. (2014), "Italy pushes ahead with 'next generation' biofuels from waste", 14 October 2014, BBC News, *www.bbc.com/news/science-environment-29618889/*.

METI (2014), Ministry of Economy Trade and Industry website, *www.meti.go.jp/english/index.html* (accessed 20 June 2015).

MINAE (2015) "Lista actualizada, Programa De Adquisición De Vehículos Eficientes Rumbo A La Carbono Neutralidad En El Transporte Primera (PAVE)" [Updated List, Acquisition Programme for Efficient Vehicles (PAVE)], 26 March 2015, Ministry of Environment of Costa Rica, *www.minae.go.cr/recursos/PAVE_2015_26_de_marzo.pdf* (accessed 20 April 2015).

Ministry of the Environment of Sweden (2014), *Sweden's Sixth National Communication on Climate Change to the United Nations Framework Convention on Climate Change*, *http://unfccc.int/files/national_reports/annex_i_natcom/submitted_natcom/application/pdf/swe_nc6_resubmission.pdf*.

NDRC (2014), *China's Policies and Actions on Climate Change 2014*, National Development and Reform Commission, *http://en.ccchina.gov.cn/archiver/ccchinaen/UpFile/Files/Default/20141126133727751798.pdf*.

OECD (2015a), *Mapping Channels to Mobilise Institutional Investment in Sustainable Energy*, Green Finance and Investment, OECD Publishing, Paris, *http://dx.doi.org/10.1787/9789264224582-en*.

OECD (2015b), *OECD Economic Surveys: United Kingdom 2015*, OECD Publishing, Paris, *http://dx.doi.org/10.1787/eco_surveys-gbr-2015-en*.

OECD (2015c), "Green Growth Indicators", *OECD Environment Statistics* (database), *http://dx.doi.org/10.1787/data-00665-en* (accessed 22 April 2015).

OECD (2015d), *OECD Economic Surveys: South Africa 2015*, OECD Publishing, Paris, *http://dx.doi.org/10.1787/eco_surveys-zaf-2015-en*.

OECD (2015e), *OECD Economic Surveys: Indonesia 2015*, OECD Publishing, Paris, *http://dx.doi.org/10.1787/eco_surveys-idn-2015-en*.

OECD (2015f), *OECD Economic Surveys: Estonia 2015*, OECD Publishing, Paris, *http://dx.doi.org/10.1787/eco_surveys-est-2015-en*.

OECD (2015g), *OECD Economic Surveys: Latvia 2015*, OECD Publishing, Paris, *http://dx.doi.org/10.1787/9789264228467-en*.

OECD (2014a), *OECD Economic Surveys: Spain 2014*, OECD Publishing, Paris, *http://dx.doi.org/10.1787/eco_surveys-esp-2014-en*.

OECD (2014b), *OECD Economic Surveys: Germany 2014*, OECD Publishing, Paris, *http://dx.doi.org/10.1787/eco_surveys-deu-2014-en*.

OECD (2014c), *OECD Economic Surveys: Poland 2014*, OECD Publishing, Paris, *http://dx.doi.org/10.1787/eco_surveys-pol-2014-en*.

OECD (2014d), "OECD ministerial statement on climate change", Meeting of the OECD Council at Ministerial Level, 6-7 May 2014, Paris, *www.oecd.org/mcm/MCM-2014-Statement-Climate-Change.pdf*.

OECD (2014e), *Green Growth Indicators for Agriculture: A Preliminary Assessment*, OECD Green Growth Studies, OECD Publishing, Paris, *http://dx.doi.org/10.1787/9789264223202-en*.

OECD (2013a), *Effective Carbon Prices*, OECD Publishing, Paris, *http://dx.doi.org/10.1787/9789264196964-en*.

OECD (2013b), "Policy approaches, by country", in *OECD Policy Instruments to Support Green Growth in Agriculture*, OECD Publishing, Paris, *http://dx.doi.org/10.1787/9789264203525-7-en*.

OECD (2013c), *OECD Economic Surveys: Brazil 2013*, OECD Publishing, Paris, *http://dx.doi.org/10.1787/eco_surveys-bra-2013-en*.

OECD (2013d), *OECD Economic Surveys: Ireland 2013*, OECD Publishing, Paris, *http://dx.doi.org/10.1787/eco_surveys-irl-2013-en*.

OECD (2012), "Greenhouse gas emissions and the potential for mitigation from materials management within OECD countries", Working Group on Waste Prevention and Recycling, OECD, Paris, *www.oecd.org/env/waste/50035102.pdf*.

Pahuja, N. et al. (2014), "GHG mitigation in India: An overview of the current policy landscape", *Working Paper*, March 2014, World Resources Institute, Washington, DC, *www.wri. org/publication/ghg-mitigation-ind-policy*.

REN21 (2015), *Renewables 2015 Global Status Report*, Renewable Energy Network for the 21st Century, Paris.

REN21 (2014), *Renewables 2014 Global Status Report*, Renewable Energy Network for the 21st Century, Paris.

Reuters (2015), *Norway to review electric car subsidies as sales soar*, *www.reuters.com/article/2015/04/20/us-norway-autos-idUSKBN0NB1T520150420* (accessed 20 July 2015).

SEIA (n.d) "The Solar Investment Tax Credit (ITC)", webpage, Solar Energy Industries Association, *www.seia.org/sites/default/files/ITC%20101%20Fact%20Sheet%20-%201-27-15.pdf* (accessed 20 June 2015).

Shiyan, C. et al. (2012), "Development of biofuels in China: Technologies, economics and policies", *Policy Research Working Papers*, 6243, World Bank Development Research Group, Washington, DC.

Smith, P. et al. (2014), "Agriculture, forestry and other land use (AFOLU)" in *Climate Change 2014: Mitigation of Climate Change. Contribution of Working Group III to the Fifth Assessment Report of the Intergovernmental Panel on Climate Change*, Edenhofer, O. et al. (eds). Cambridge University Press, Cambridge, UK and New York, US.

UNEP (2015), *The Chinese Automotive Fuel Economy Policy*, *www.unep.org/transport/gfei/autotool/case_studies/apacific/china/CHINA%20CASE%20STUDY.pdf* (accessed 25 August 2015).

UNFCCC (2015a), "INDCs as communicated by Parties", web portal, United Nations Framework Convention on Climate Change, *www.4.unfccc.int/submissions/indc/Submission%20Pages/submissions.aspx* (accessed 20 July 2015).

UNFCCC (2015b), "Submissions from Parties on proposed forest reference emission levels and/or forest reference levels for the implementation of the activities referred to in decision 1/CP.16, paragraph 70", webpage, United Nations Framework Convention on Climate Change, *http://unfccc.int/land_use_and_climate_change/redd/items/8414.php* (accessed 08 July 2015).

UNFCCC (2013), "Compilation of information on nationally appropriate mitigation actions to be implemented by developing country parties", United Nations Framework Convention on Climate Change, *http://unfccc.int/resource/docs/2013/sbi/eng/inf12r02.pdf*.

US Department of State (2014), *United States Climate Action Report 2014*, United States Department of State, *www.state.gov/documents/organization/219038.pdf*.

Wreford, A., D. Moran and N. Adger (2010), *Climate Change and Agriculture: Impacts, Adaptation and Mitigation*, OECD Publishing, Paris, *http://dx.doi.org/10.1787/9789264086876-en*.

ANNEX A

Individual country analysis: Methodology and results

Annual rates of change of GHG emissions and emissions intensity

Table 2.5 shows historical annual rates of change of GHG emissions or emission intensity, and compares them with the future average rates of change needed to meet the mitigation targets and goals that have been announced. For all countries except China and India, historical rates of change of total GHG emissions are calculated on a compound basis for three periods: (i) 1990-2000; (ii) 2000-05; and (iii) 2005-12. The maximum and minimum values observed since 1990 are also shown, using a three-year rolling average. For China and India, historical rates of change of CO_2 intensity (i.e. CO_2 emissions per unit GDP) are shown. Note that India's 2020 pledge is for "emissions intensity" excluding emissions from the agriculture sector.

In most cases, national GHG inventory statistics are used. In some cases, the historical rates of change are not available because only partial national GHG statistics are available. For China and India, IEA statistics on CO_2 emissions from fossil fuel combustion are used, since the latest national GHG inventories available are 2005 and 2000 respectively. The GDP data used are from *OECD National Accounts* for OECD member countries and the World Bank for partner economies, calculated in 2005 USD using the purchasing power parity approach and available via the *IEA CO_2 Emissions from Fuel Combustion Statistics* database. In cases where LULUCF is included, this is done by calculating a percentage reduction from GHG emissions including LULUCF in the base year. It should be recognised that this is a simplified approximation of LULUCF accounting and that in reality LULUCF accounting is more complicated.

In the case of absolute emission reduction targets for 2020, future annual rates of change of GHG emissions are calculated on a compound basis between the latest year for which data are available (2012 for most countries) and 2020. For emission reduction targets for 2025 and 2030, annual rates of change are calculated for the period 2020-25 or 2020-30. Where a range of targets has been submitted, a range of rates is calculated. For countries that have pledged a range of targets for 2020, as well as a target for 2030, post-2020 rates of change are calculated using both the high and low ends of the 2020 target.

The statistical data for Israel are supplied by and under the responsibility of the relevant Israeli authorities. The use of such data by the OECD is without prejudice to the status of the Golan Heights, East Jerusalem and Israeli settlements in the West Bank under the terms of international law.

In the case of the EU-28, EU Member States will jointly fulfil their 2020 and 2030 emission reduction targets. Therefore, individual EU Member States are not covered individually in this analysis.

In the case of countries that have pledged to reduce GHG emissions from BAU baselines, the following sources are used for the BAU baseline:

- **Brazil:** *Estimativas anuais de emissões de gases de efeito estufa - 2ª edição* (2014).[1]
- **Chile:** Medium GDP growth emission projection from *Mitigation Options for a Low Carbon Development*, MAPS Chile, Phase 2.[2]
- **Indonesia:** No national BAU baseline data available.
- **Israel:** Emissions baseline from the Heifetz and McKinsey study (2009), as referenced in the *OECD Environmental Performance Reviews: Israel 2011*.[3]
- **Korea:** Emissions baseline from Korea's intended nationally determined contribution submitted to the UNFCCC.[4]
- **Mexico:** Emissions baseline contained in Mexico's INDC submission.[5]
- **South Africa:** "Growth Without Constraints" baseline from South Africa's *Long-Term Mitigation Scenarios* project.[6]

Rates of change are not analysed for Costa Rica's pledge to become carbon neutral by 2021 because the methodology is not well suited to an analysis of near-zero emissions pathways.

Emission intensity and GDP scatter plots

The starting point for this part of the analysis is to consider the key drivers behind GHG emissions. Total GHG emissions in year n are the product of two drivers: (i) GDP (GDP_n, in USD); and (ii) GHG emissions intensity per unit of GDP (I_n, in $kgCO_2e$ per unit of GDP):

$$GHG_n = GDP_n \times I_n$$

For each country, the historical rates of change of GDP and emissions intensity are calculated for the period 1990-2012 (or the latest year for which data are available). Different colours distinguish between data points from 1990-2000 and 2001-12. To smooth out short-term fluctuations, three-year rolling compound averages for these annual growth rates are used where full-time series are available.

The rates of change of emission intensity and GDP needed to meet the targets and goals that have been announced are also calculated. A higher GDP growth rate implies that a faster average decrease in emissions intensity will be needed to meet a given emissions target or goal. The result is therefore a line or frontier of values marking the point at which the target or goal would be met. Beyond the frontier, countries would exceed their targeted emissions levels.

Individual country results are presented below. Average rates of change of emission intensity and GDP would need to be in the area below the lines marked on the scatter plots in order to meet the mitigation targets and goals indicated. There are some important caveats to this simple analysis. First, full-time series of GHG statistics are not available for all countries. Second, most of the analyses exclude emissions and removals from the LULUCF sector since it remains unclear how the LULUCF sector will be accounted for in mitigation pledges and INDCs. Third, use of offsets from market mechanisms is excluded. In reality, several countries intend to use carbon credits from international market mechanisms to meet part of their mitigation targets and goals. This would decrease the rates of domestic emission reduction required to meet any given mitigation target or goal.

Figure A.1. **Individual country analysis: Australia**

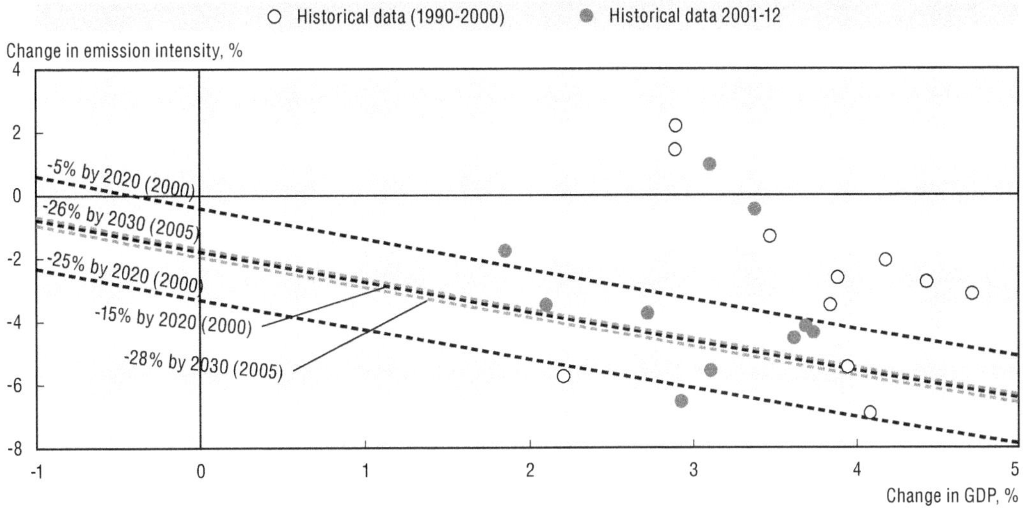

1. Includes LULUCF. The rates of change for 2020-30 are calculated assuming that a 5% reduction by 2020 from 2000 levels is achieved.

Source: GHG statistics from Government of Australia (2015), *National Inventory Report 2013*, *http://unfccc.int/national_reports/annex_i_ghg_inventories/national_inventories_submissions/items/8812.php* (accessed 25 August 2015); GDP statistics from IEA (2015), "Indicators for CO_2 emissions", *IEA CO_2 Emissions from Fuel Combustion Statistics* (database), *http://dx.doi.org/10.1787/data-00433-en* (accessed 21 July 2015).

StatLink http://dx.doi.org/10.1787/888933272543

Figure A.2. **Individual country analysis: Brazil**

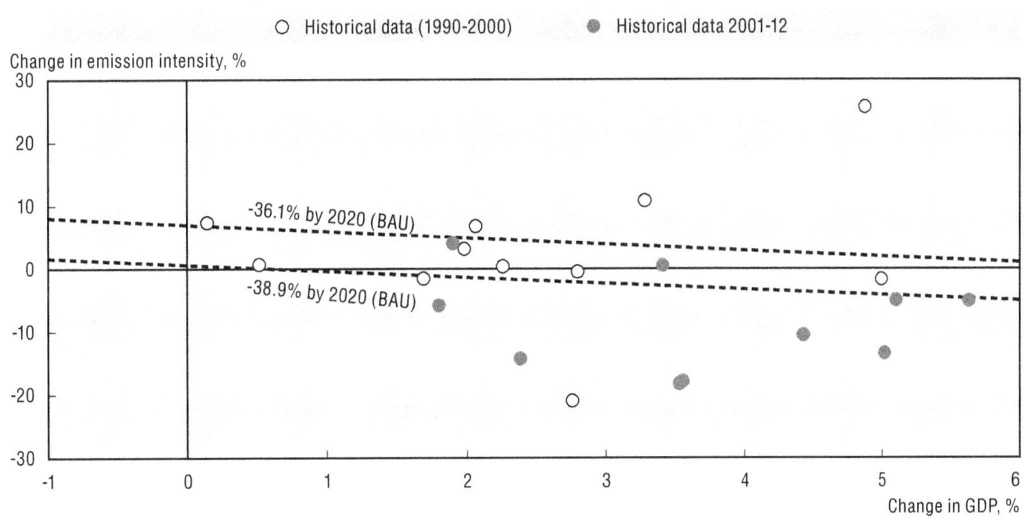

1. Includes LULUCF to match the scope of Brazil's mitigation pledge (which includes LULUCF). Based on national BAU baseline from Government of Brazil (2014).

Source: GHG statistics and BAU emissions data from Government of Brazil (2014) *Estimativas anuais de emissões de gases de efeito estufa - 2ª edição* [Annual estimations of greenhouse gas emissions – second editions], *www.mct.gov.br/upd_blob/0235/235580.pdf* (in Portuguese); GDP statistics from IEA (2015), "Indicators for CO_2 emissions", *IEA CO_2 Emissions from Fuel Combustion Statistics* (database), *http://dx.doi.org/10.1787/data-00433-en* (accessed 21 July 2015); OECD (2015) *OECD Environmental Performance Reviews: Brazil 2015*.

StatLink http://dx.doi.org/10.1787/888933272558

Figure A.3. **Individual country analysis: Canada**

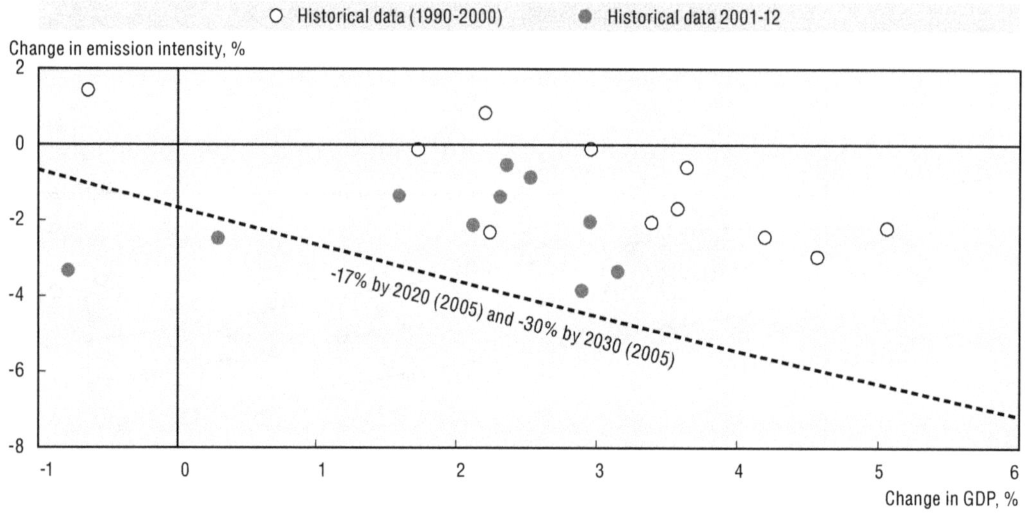

1. Excluding LULUCF.

Source: GDP statistics from IEA (2015), "Indicators for CO_2 emissions", *IEA CO_2 Emissions from Fuel Combustion Statistics* (database), *http://dx.doi.org/10.1787/data-00433-en* (accessed 21 July 2015); GHG statistics from UNFCCC (2015), GHG inventory data detailed by Party, http://unfccc.int/di/DetailedByParty.do (accessed 07 July 2015).

StatLink *http://dx.doi.org/10.1787/888933272568*

Figure A.4. **Individual country analysis: Chile**

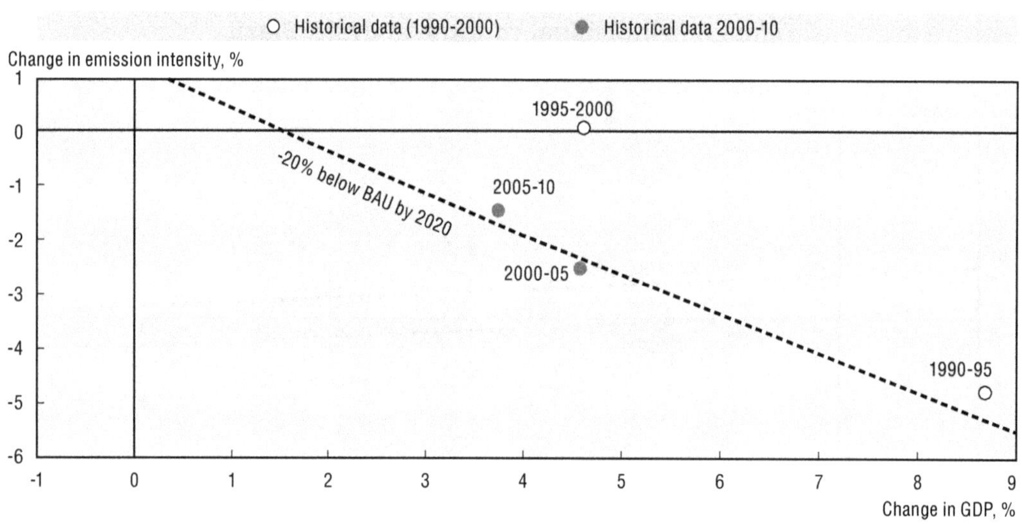

1. Excludes LULUCF. Based on Chile's BAU baseline from the MAPS project, Phase 2.

Source: MAPS Chile (2014), Mitigation Options for a Low Carbon Development, www.mapsprogramme.org/wp-content/uploads/Chile-Phase-2-Synthesis-of-Results_English_Final.pdf; GDP statistics from IEA (2015), "Indicators for CO2 emissions", IEA CO2 Emissions from Fuel Combustion Statistics (database), http://dx.doi.org/10.1787/data-00433-en (accessed 21 July 2015); GHG statistics from UNFCCC (2015), GHG inventory data detailed by Party. http://unfccc.int/di/DetailedByParty.do (accessed 07 July 2015).

StatLink *http://dx.doi.org/10.1787/888933272572*

Figure A.5. **Individual country analysis: China**

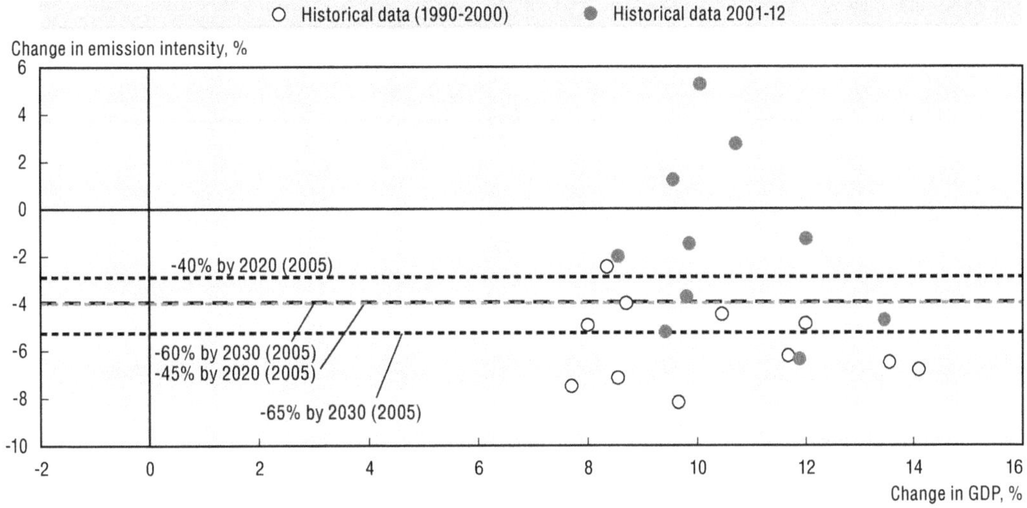

1. Based on IEA statistics for CO_2 from fossil fuel combustion to match the scope of China's mitigation pledge (which is for CO_2 intensity only).

Source: CO_2 and GDP statistics from IEA (2015), "CO_2 emissions by product and flow", *IEA CO_2 Emissions from Fuel Combustion Statistics* (database), *http://dx.doi.org/10.1787/data-00430-en* (accessed 21 July 2015); IEA (2015), "Indicators for CO_2 emissions", *IEA CO_2 Emissions from Fuel Combustion Statistics* (database), *http://dx.doi.org/10.1787/data-00433-en* (accessed 21 July 2015).

StatLink http://dx.doi.org/10.1787/888933272581

Figure A.6. **Individual country analysis: Colombia**

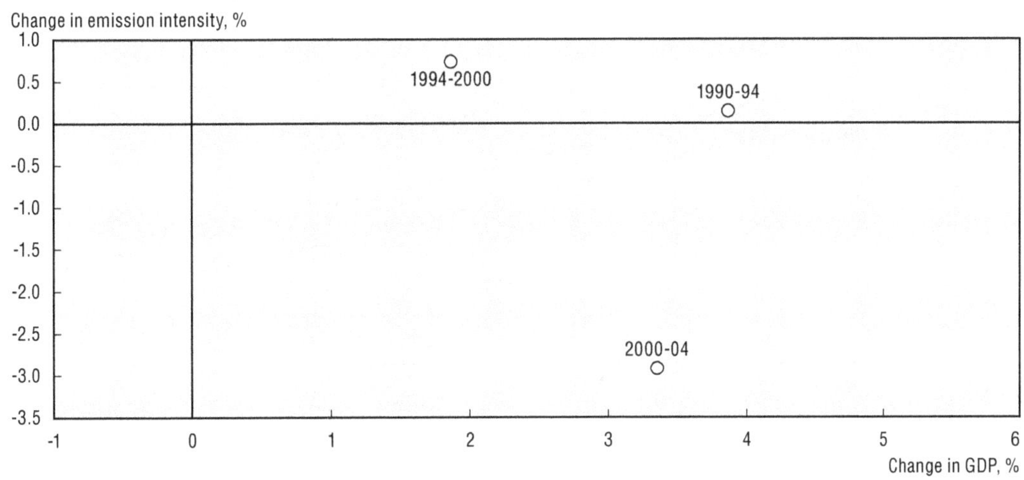

1. Includes LULUCF. Lines are not shown for Colombia's 2020 mitigation pledges because they do not include an emissions reduction target.

Source: GDP statistics from IEA (2015), "Indicators for CO_2 emissions", *IEA CO_2 Emissions from Fuel Combustion Statistics* (database), *http://dx.doi.org/10.1787/data-00433-en* (accessed 21 July 2015); GHG statistics from UNFCCC (2015), GHG inventory data detailed by Party, http://unfccc.int/di/DetailedByParty.do (accessed 07 July 2015).

StatLink http://dx.doi.org/10.1787/888933272594

Figure A.7. **Individual country analysis: Costa Rica**

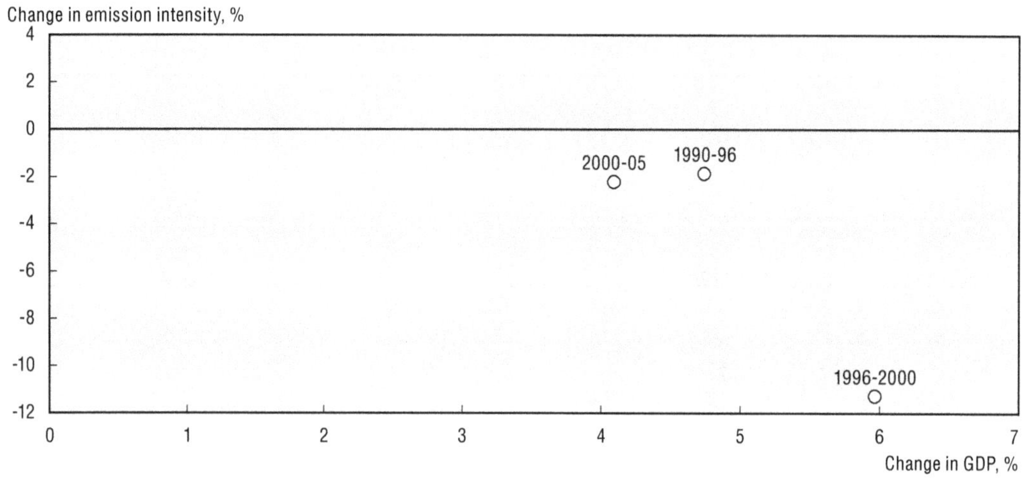

1. Includes LULUCF. Lines are not shown for Costa Rica's 2021 carbon neutrality goal.

Source: GDP statistics from IEA (2015), "Indicators for CO_2 emissions", *IEA CO_2 Emissions from Fuel Combustion Statistics* (database), *http://dx.doi.org/10.1787/data-00433-en* (accessed 21 July 2015); GHG statistics from UNFCCC (2015), GHG inventory data detailed by Party, http://unfccc.int/di/DetailedByParty.do (accessed 7 July 2015).

StatLink *http://dx.doi.org/10.1787/888933272604*

Figure A.8. **Individual country analysis: EU-28**

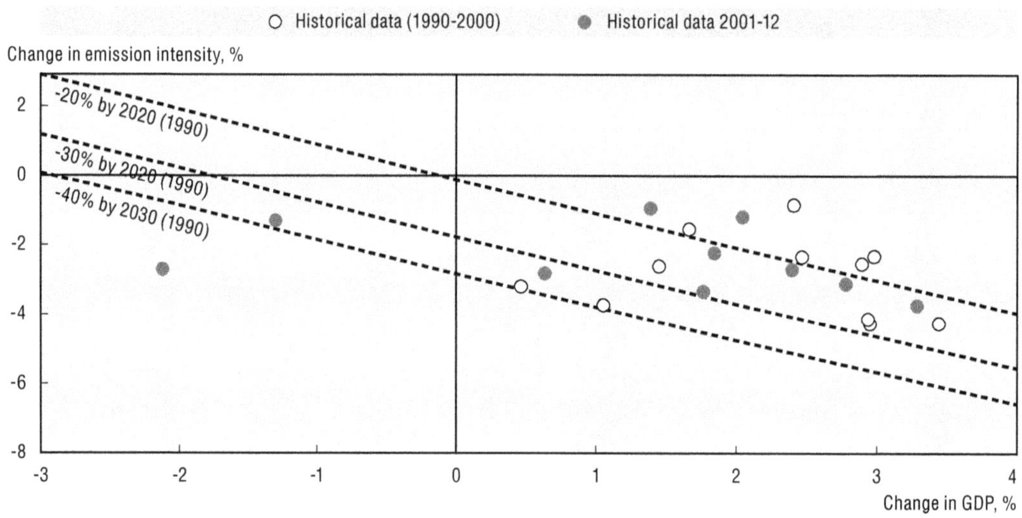

1. Excluding LULUCF. The rates of change for 2020-30 are calculated assuming that the low end of the 2020 pledge is achieved (i.e. –20% by 2020).

Source: GDP statistics from IEA (2015), "Indicators for CO_2 emissions", *IEA CO_2 Emissions from Fuel Combustion Statistics* (database), *http://dx.doi.org/10.1787/data-00433-en* (accessed 21 July 2015); GHG statistics from UNFCCC (2015), GHG inventory data detailed by Party, http://unfccc.int/di/DetailedByParty.do (accessed 07 July 2015).

StatLink *http://dx.doi.org/10.1787/888933272615*

Figure A.9. **Individual country analysis: Iceland**

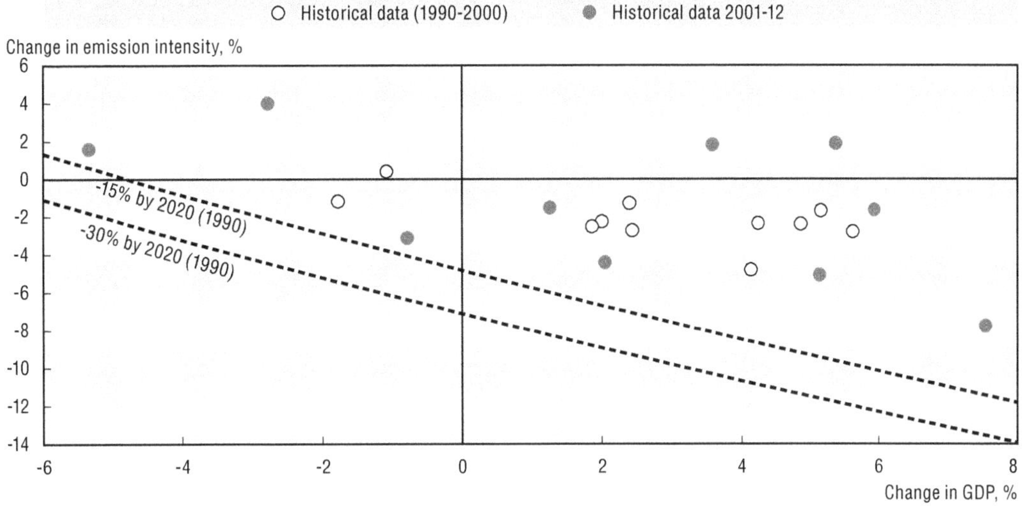

1. Excludes LULUCF. For 2030, Iceland aims to be part of a collective delivery by European countries to reach a target of 40% reduction of greenhouse gas emissions compared to 1990 levels.

Source: GDP statistics from IEA (2015), "Indicators for CO_2 emissions", *IEA CO_2 Emissions from Fuel Combustion Statistics* (database), *http://dx.doi.org/10.1787/data-00433-en* (accessed 21 July 2015); GHG statistics from UNFCCC (2015), GHG inventory data detailed by Party, http://unfccc.int/di/DetailedByParty.do (accessed 07 July 2015).

StatLink http://dx.doi.org/10.1787/888933272625

Figure A.10. **Individual country analysis: India**

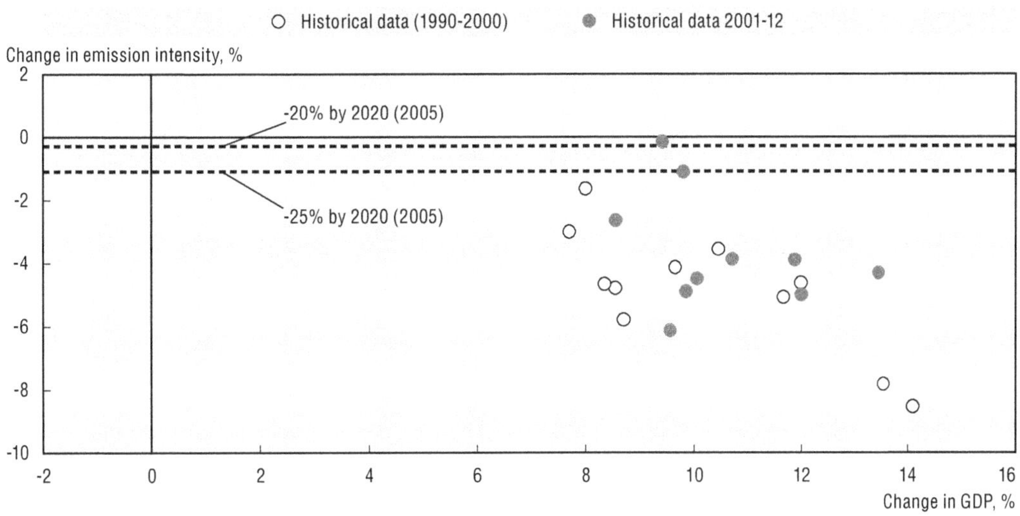

1. Based on IEA statistics for CO_2 from fossil fuel combustion. Note that India's pledge excludes emissions from agriculture.

Source: CO_2 and GDP statistics from IEA (2015), "CO_2 emissions by product and flow", *IEA CO_2 Emissions from Fuel Combustion Statistics* (database), *http://dx.doi.org/10.1787/data-00430-en* (accessed 21 July 2015); GDP statistics from IEA (2015), "Indicators for CO_2 emissions", *IEA CO_2 Emissions from Fuel Combustion Statistics* (database), *http://dx.doi.org/10.1787/data-00433-en* (accessed 21 July 2015).

StatLink http://dx.doi.org/10.1787/888933272638

Figure A.11. **Individual country analysis: Indonesia**

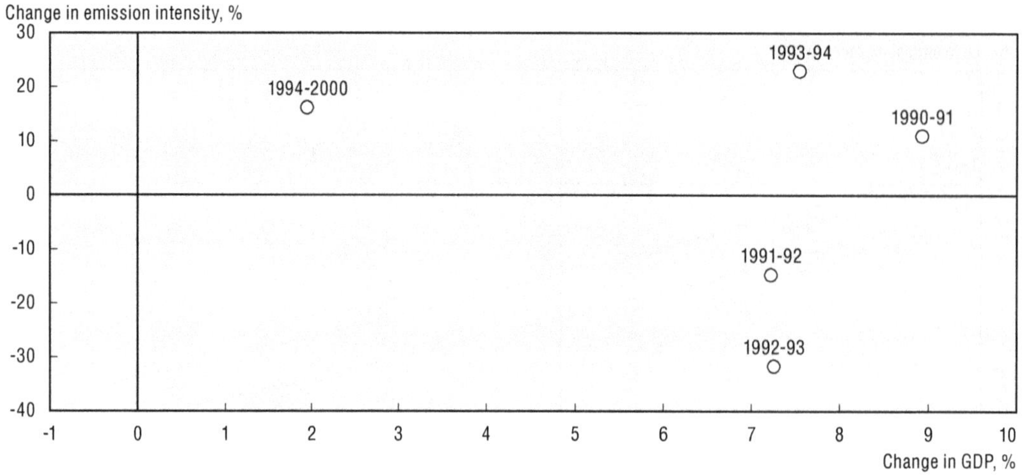

1. Includes LULUCF. Lines are not shown for Indonesia's 2020 mitigation pledge because information on Indonesia's BAU baseline is unavailable.

Source: GDP statistics from IEA (2015), "Indicators for CO_2 emissions", *IEA CO_2 Emissions from Fuel Combustion Statistics* (database), *http://dx.doi.org/10.1787/data-00433-en* (accessed 21 July 2015); GHG statistics from UNFCCC (2015), GHG inventory data detailed by Party, *http://unfccc.int/di/DetailedByParty.do* (accessed 07 July 2015).

StatLink *http://dx.doi.org/10.1787/888933272641*

Figure A.12. **Individual country analysis: Israel**

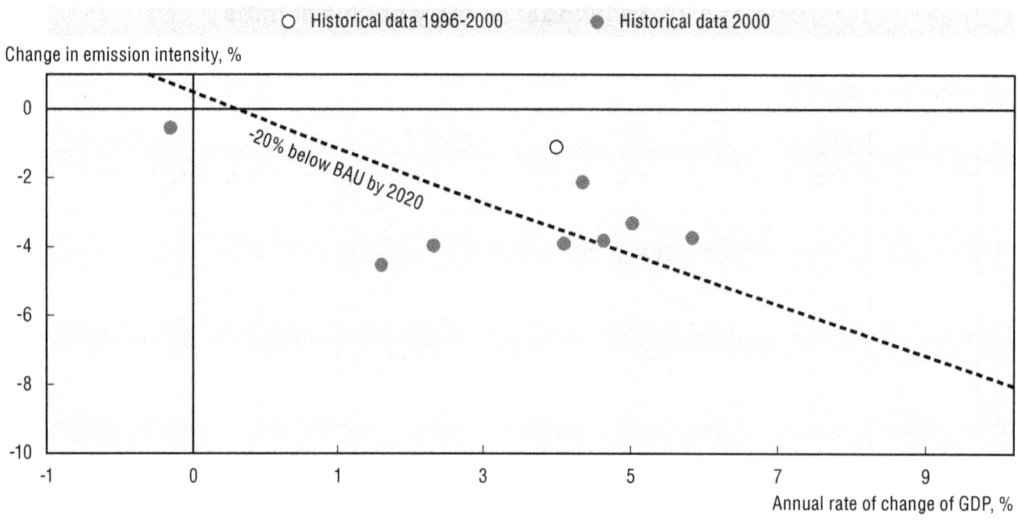

1. Excludes LULUCF. Uses the study of Heifetz and Mc Kinsey for the BAU baseline information.

Source: OECD (2011), *OECD Environmental Performance Reviews: Israel 2011*, OECD Publishing, Paris, *http://dx.doi.org/10.1787/9789264117563-en*; GDP statistics from IEA (2014); GHG statistics from UNFCCC (2015), GHG inventory data detailed by Party. http://unfccc.int/di/DetailedByParty.do (accessed 07 July 2015).

StatLink *http://dx.doi.org/10.1787/888933272652*

Figure A.13. **Individual country analysis: Japan**

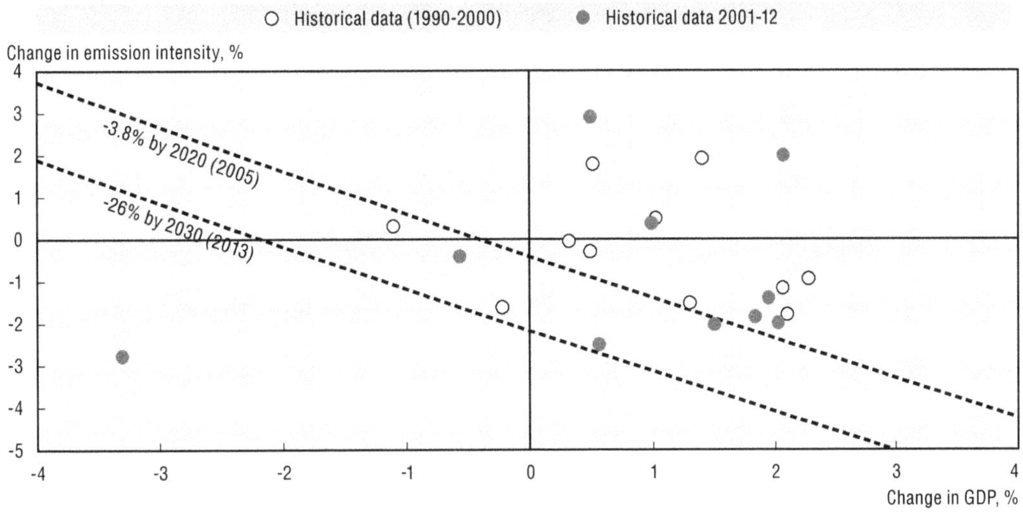

1. Excluding LULUCF.

Source: GDP statistics from IEA (2015), "Indicators for CO_2 emissions", *IEA CO_2 Emissions from Fuel Combustion Statistics* (database), *http://dx.doi.org/10.1787/data-00433-en* (accessed 21 July 2015); GHG statistics from UNFCCC (2015), GHG inventory data detailed by Party. *http://unfccc.int/di/DetailedByParty.do* (accessed 07 July 2015).

StatLink http://dx.doi.org/10.1787/888933272660

Figure A.14. **Individual country analysis: Korea**

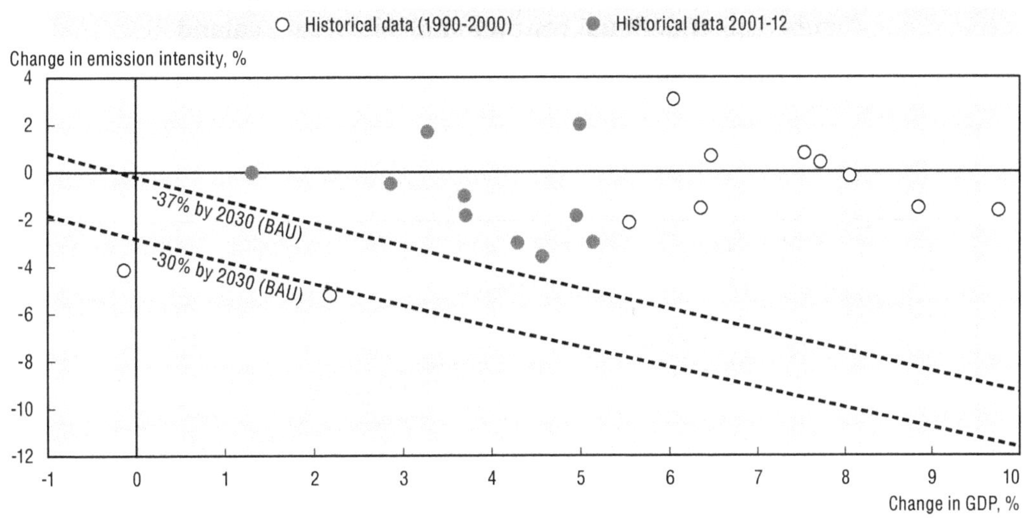

1. Excluding LULUCF. Based on BAU baseline from Korea's INDC submission to the UNFCCC.

Source: GHG statistics from Government of Korea (2015), *Korea's Intended Nationally Determined Contribution, www.4.unfccc.int/submissions/INDC/* (accessed 21 July 2015) and UNFCCC (2015), GHG inventory data detailed by Party. *http://unfccc.int/di/DetailedByParty.do* (accessed 07 July 2015); GDP statistics from IEA (2015), "Indicators for CO_2 emissions", *IEA CO_2 Emissions from Fuel Combustion Statistics* (database), *http://dx.doi.org/10.1787/data-00433-en* (accessed 21 July 2015).

StatLink http://dx.doi.org/10.1787/888933272670

Figure A.15. **Individual country analysis: Mexico**

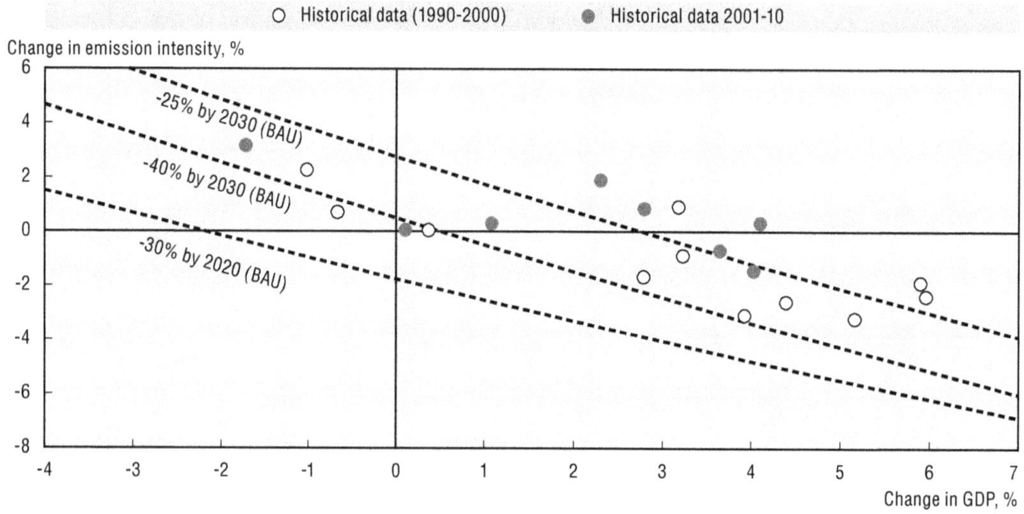

1. Excludes LULUCF. Uses national BAU baseline information from Mexico's INDC submission.

Source: GHG statistics from Government of Mexico (2015), *Mexico's Intended Nationally Determined Contribution*, *www.4.unfccc.int/submissions/INDC/Published%20Documents/Mexico/1/MEXICO%20INDC%2003.30.2015.pdf* (accessed 21 July 2015) and UNFCCC (2015), GHG inventory data detailed by Party, *http://unfccc.int/di/DetailedByParty.do* (accessed 07 July 2015); GDP statistics from IEA (2015), "Indicators for CO_2 emissions", *IEA CO_2 Emissions from Fuel Combustion Statistics* (database), *http://dx.doi.org/10.1787/data-00433-en* (accessed 21 July 2015).

StatLink *http://dx.doi.org/10.1787/888933272682*

Figure A.16. **Individual country analysis: New Zealand**

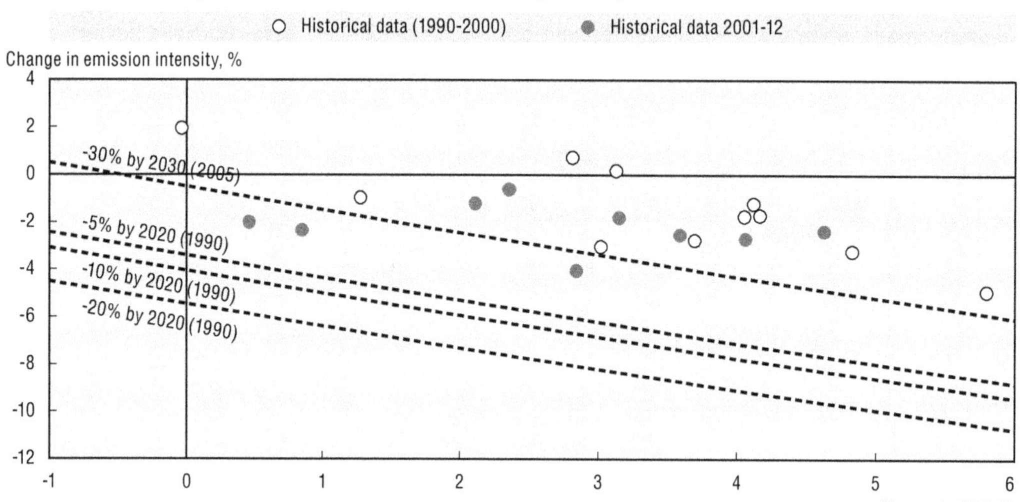

1. Excludes LULUCF. The rates of change for 2020-30 are calculated assuming that the low end of the 2020 pledge is achieved (i.e. –5% by 2020).

Source: GDP statistics from IEA (2015), "Indicators for CO_2 emissions", *IEA CO_2 Emissions from Fuel Combustion Statistics* (database), *http://dx.doi.org/10.1787/data-00433-en* (accessed 21 July 2015); GHG statistics from UNFCCC (2015), GHG inventory data detailed by Party, *http://unfccc.int/di/DetailedByParty.do* (accessed 07 July 2015).

StatLink *http://dx.doi.org/10.1787/888933272695*

Figure A.17. **Individual country analysis: Norway**

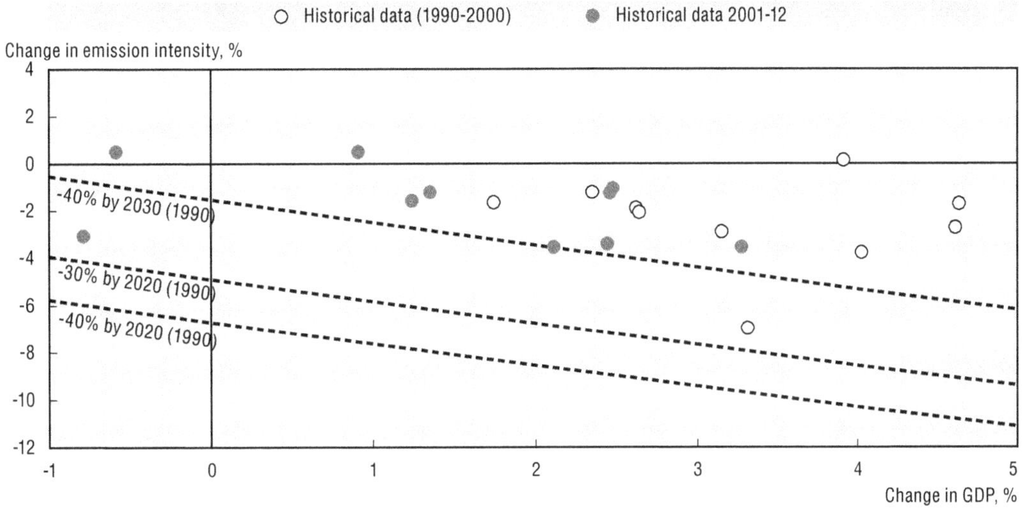

1. Excludes LULUCF. For 2030, Norway intends to fulfil its INDC through a collective delivery with the EU and its Member States. In the event that there is no agreement on a collective delivery with the EU, Norway will fulfil the commitment individually. In this situation, Norway assumes that it will have access to flexible mechanisms as in the case with collective delivery with the EU. The ambition level will remain the same in this event.

Source: GDP statistics from IEA (2015), "Indicators for CO_2 emissions", *IEA CO_2 Emissions from Fuel Combustion Statistics* (database), *http://dx.doi.org/10.1787/data-00433-en* (accessed 21 July 2015); GHG statistics from UNFCCC (2015), GHG inventory data detailed by Party, *http://unfccc.int/di/DetailedByParty.do* (accessed 07 July 2015).

StatLink http://dx.doi.org/10.1787/888933272708

Figure A.18. **Individual country analysis: Russian Federation**

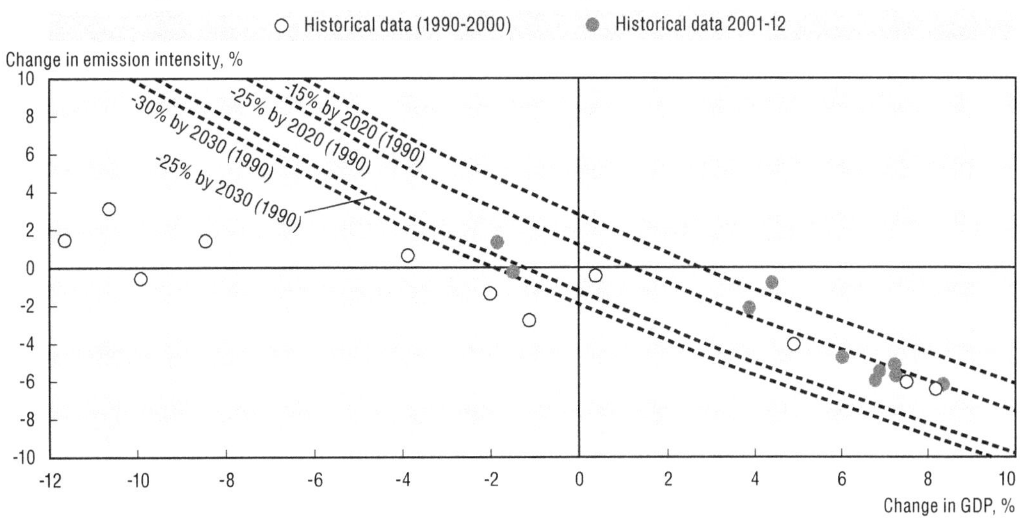

1. Excludes LULUCF. The rates of change for 2020-30 are calculated assuming that the low end of the 2020 pledge is achieved (i.e. –15% by 2020).

Source: GDP statistics from IEA (2015), "Indicators for CO_2 emissions", *IEA CO_2 Emissions from Fuel Combustion Statistics* (database), *http://dx.doi.org/10.1787/data-00433-en* (accessed 21 July 2015); GHG statistics from UNFCCC (2015), GHG inventory data detailed by Party, *http://unfccc.int/di/DetailedByParty.do* (accessed 07 July 2015).

StatLink http://dx.doi.org/10.1787/888933272719

Figure A.19. **Individual country analysis: South Africa**

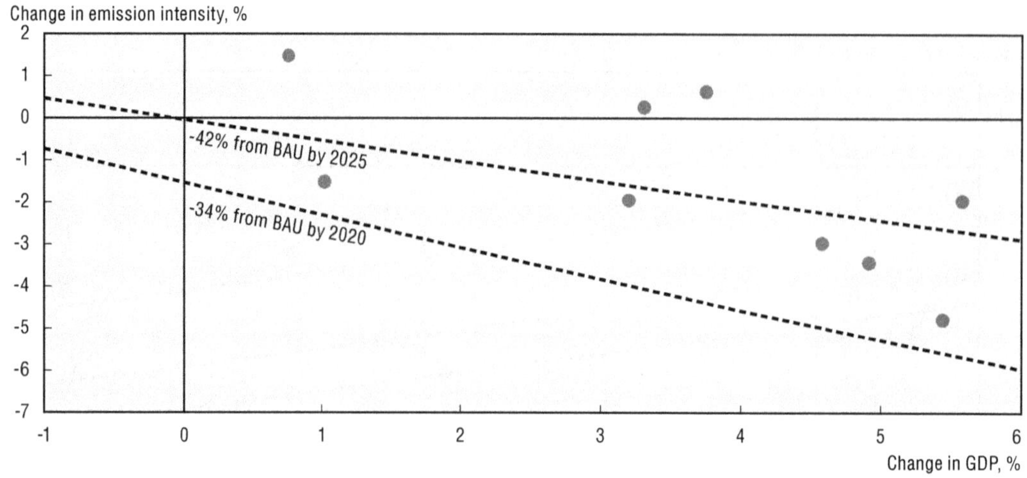

1. Excludes LULUCF. Uses the "Growth Without Constraints" emissions baseline from South Africa (2007).

Source: Government of South Africa (2007), *Long Term Mitigation Scenarios: Strategic Options for South Africa*, *www.erc.uct.ac.za/Research/publications/07Scenario_team-LTMS_Scenarios.pdf*; GDP statistics from IEA (2015), "Indicators for CO_2 emissions", *IEA CO_2 Emissions from Fuel Combustion Statistics* (database), *http://dx.doi.org/10.1787/data-00433-en* (accessed 21 July 2015); GHG statistics from UNFCCC (2015), GHG inventory data detailed by Party, http://unfccc.int/di/DetailedByParty.do (accessed 07 July 2015).

StatLink http://dx.doi.org/10.1787/888933272725

Figure A.20. **Individual country analysis: Switzerland**

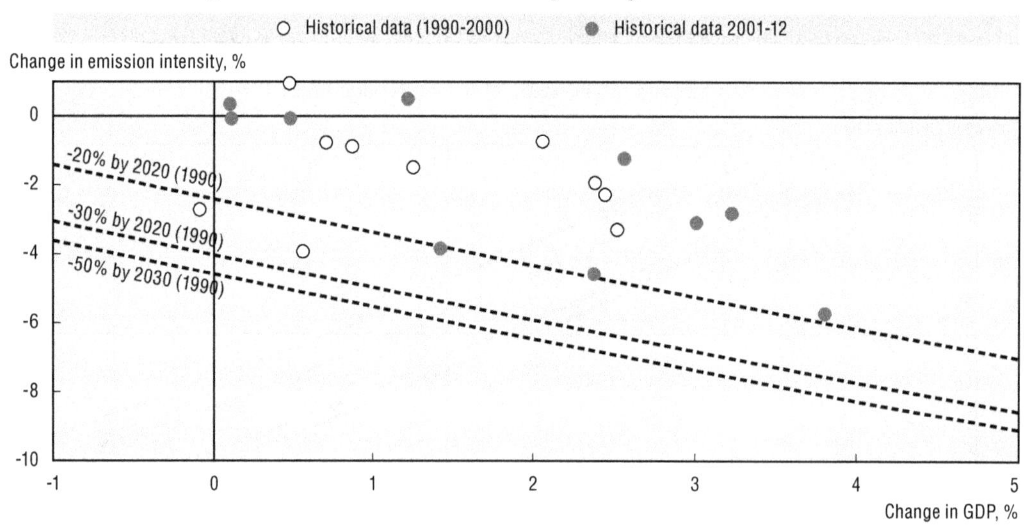

1. Excludes LULUCF. The rates of change for 2020-30 are calculated assuming that the low end of the 2020 pledge is achieved (i.e. –20% by 2020).

Source: GDP statistics from IEA (2015), "Indicators for CO_2 emissions", *IEA CO_2 Emissions from Fuel Combustion Statistics* (database), *http://dx.doi.org/10.1787/data-00433-en* (accessed 21 July 2015); GHG statistics from UNFCCC (2015), GHG inventory data detailed by Party, *http://unfccc.int/di/DetailedByParty.do* (accessed 07 July 2015).

StatLink http://dx.doi.org/10.1787/888933272739

Figure A.21. **Individual country analysis: Turkey**

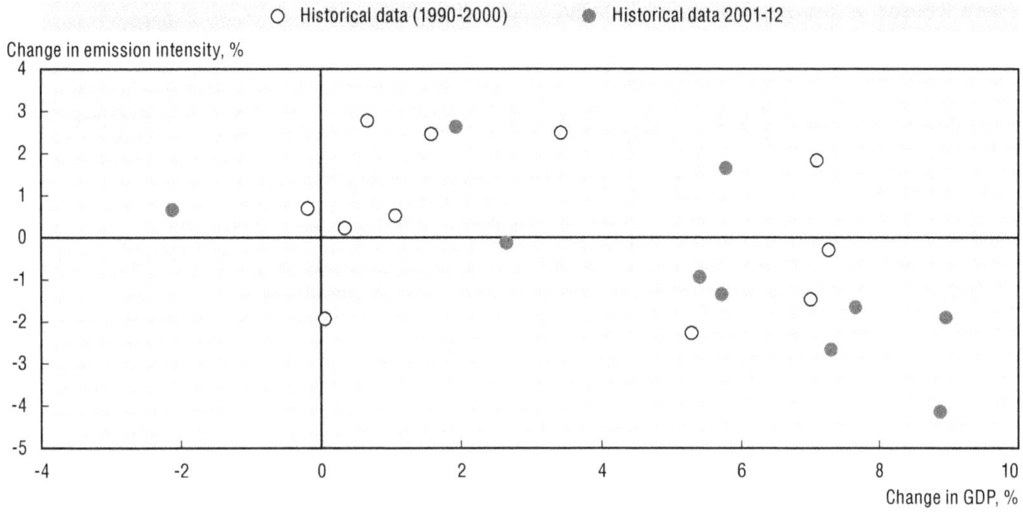

1. Excluding LULUCF.

Source: GDP statistics from IEA (2015), "Indicators for CO_2 emissions", *IEA CO_2 Emissions from Fuel Combustion Statistics* (database), *http://dx.doi.org/10.1787/data-00433-en* (accessed 21 July 2015); GHG statistics from UNFCCC (2015), GHG inventory data detailed by Party, *http://unfccc.int/di/DetailedByParty.do* (accessed 07 July 2015).

StatLink http://dx.doi.org/10.1787/888933272746

Figure A.22. **Individual country analysis: United States**

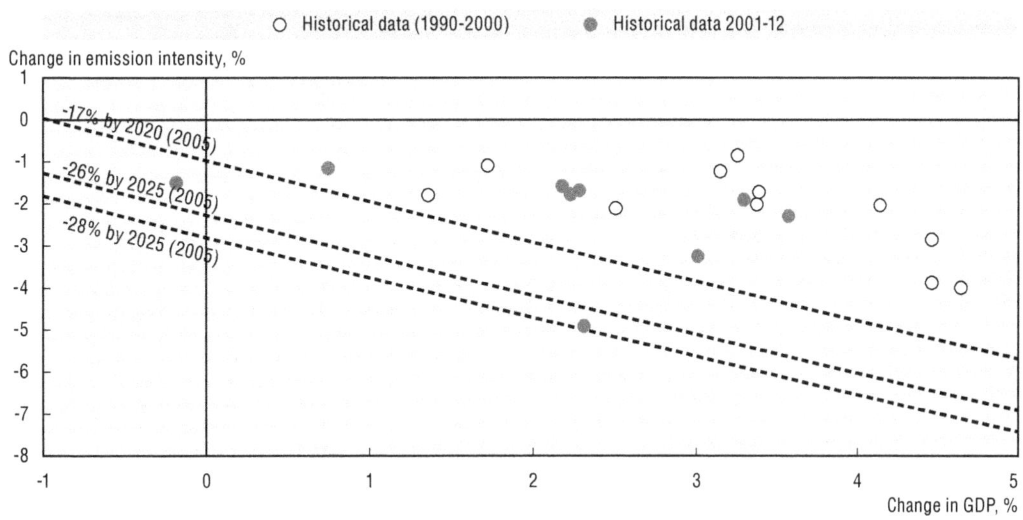

1. Excluding LULUCF.

Source: GDP statistics from IEA (2015), "Indicators for CO_2 emissions", *IEA CO_2 Emissions from Fuel Combustion Statistics* (database), *http://dx.doi.org/10.1787/data-00433-en* (accessed 21 July 2015); GHG statistics from UNFCCC (2015), GHG inventory data detailed by Party, *http://unfccc.int/di/DetailedByParty.do* (accessed 07 July 2015).

StatLink http://dx.doi.org/10.1787/888933272756

Notes

1. Available at: *www.mct.gov.br/upd_blob/0235/235580.pdf* (in Portuguese).
2. Available at: *www.mapsprogramme.org/wp-content/uploads/Chile-Phase-2-Synthesis-of-Results_English_Final.pdf*.
3. OECD (2011), *OECD Environmental Performance Reviews: Israel 2011*, OECD Publishing, Paris, *http://dx.doi.org/10.1787/9789264117563-en*.
4. Available at: *www4.unfccc.int/submissions/indc/Submission%20Pages/submissions.aspx*.
5. Available at: *www4.unfccc.int/submissions/indc/Submission%20Pages/submissions.aspx*.
6. Government of South Africa (2007), *Long-Term Mitigation Scenarios: Strategic Options for South Africa*, *www.erc.uct.ac.za/Research/publications/07Scenario_team-LTMS_Scenarios.pdf*.

Glossary

This glossary draws on definitions from the OECD[1] and the International Energy Agency, as well as other sources such as the Intergovernmental Panel on Climate Change, the United Nations Framework Convention on Climate Change and REN21.

International negotiations

- **Annex I Parties** include industrialised countries that were members of the OECD in 1992, plus countries with economies in transition (including the Russian Federation, the Baltic States, and several Central and Eastern European States).
- **Carbon credits or offsets** refer to the use of credits to meet a mitigation goal that represent emission reductions or removals achieved elsewhere. For the first and second commitment periods of the Kyoto Protocol, there are clear accounting rules regarding the use of GHG units to meet commitments. However, for mitigation pledges for 2020 under the UNFCCC and post-2020 Intended Nationally Determined Contributions, there are currently no agreed accounting rules regarding the use of GHG units. Offsets do not affect the level of emissions at a global level.
- **Intended Nationally Determined Contributions (INDCs)** are pledges that countries are making ahead of the COP 21 climate change conference in Paris in 2015. All governments have been invited to communicate their INDCs by the first quarter of 2015 (if ready to do so). INDCs are to contribute towards climate change mitigation, and may also include an adaptation component. INDCs are to be communicated in a manner that facilitates the clarity, transparency and understanding of the INDC.
- **Kyoto Protocol** is an international agreement linked to the UNFCCC, which commits its Parties by setting internationally binding emission reduction targets. The Kyoto Protocol was adopted in Kyoto, Japan, on 11 December 1997 and entered into force on 16 February 2005. The detailed rules for the implementation of the protocol were adopted at COP 7 in Marrakesh, Morocco, in 2001, and are referred to as the "Marrakesh Accords." Its first commitment period started in 2008 and ended in 2012. Its second started in 2013 and ends in 2020.
- **Land use, land-use change and forestry (LULUCF)** refers to activities that contribute to abating emissions by increasing the removals of greenhouse gases from the atmosphere (e.g. by planting trees or managing forests), or by reducing emissions (e.g. by curbing deforestation). However, there are drawbacks as it may often be difficult to estimate greenhouse gas removals and emissions resulting from these activities. The 1996 IPCC Guidelines for National Greenhouse Gas Inventories provides approaches and methodology to account for the LULUCF category in national GHG registries. Under the Kyoto Protocol, Parties decided that greenhouse gas removals and emissions through certain activities — namely, afforestation and reforestation since 1990 — are accounted for in meeting the Kyoto Protocol's emission targets.
- **Nationally Appropriate Mitigation Action (NAMA)** are economy-wide or sector-specific actions agreed by developing country Parties under the UNFCCC.

- **Non-Annex I Parties** are mostly developing countries. Certain groups of developing countries are recognised by the convention as being especially vulnerable to the adverse impacts of climate change. Others (such as countries that rely heavily on income from fossil fuel production and commerce) feel more vulnerable to the potential economic impacts of climate change response measures. The convention emphasises activities that promise to answer the special needs and concerns of these vulnerable countries, such as investment, insurance and technology transfer.
- **United Nations Framework Convention on Climate Change (UNFCC)** is an international convention that was adopted in 1992 and entered in force in 1994. There are 196 Parties to the convention. The objective of the treaty is to stabilise greenhouse gas concentrations in the atmosphere at a level that would prevent dangerous anthropogenic interference with the climate system.

Energy

- **Carbon capture and storage (CCS)** involves the use of technology, first to collect and concentrate the CO_2 produced in industrial and energy-related sources, transport it to a suitable storage location, and then store it away from the atmosphere for a long period of time. CCS would thus allow fossil fuels to be used with low emissions of greenhouse gases.
- **Energy supply sector**, as defined by the IPCC guidelines for GHG inventories, comprises energy extraction, conversion, storage, transmission and distribution processes that deliver final energy to the end-use sectors (industry, transport and building, as well as agriculture and forestry). In some parts of this report, transport is separated from the energy sector in the context of domestic policy making.
- **Low-carbon technologies** include renewable energy sources, nuclear power and CCS technologies.
- **Research, development and demonstration (RD&D)** refers to scientific and/or technical research and development of new production processes or products, coupled with analysis and measures that provide information to potential users regarding the application of the new product or process; demonstration tests; and feasibility of applying these product processes via pilot plants and other pre-commercial applications.
- **Renewable energy** refers to energy sourced from natural processes that are continuously replenished. These include solar, wind, geothermal, hydro, tide, wave, the renewable component of waste and biofuels. Most of the use of renewable energy is free of CO_2.
- **Renewable energy target** is an official commitment, plan or goal set by a government (at the local, state, national or regional level) to achieve a certain amount of renewable energy by a future date. Some targets are legislated, while others are set by regulatory agencies or ministries.
- **Smart grids** use information and communications technology to co-ordinate the needs and capabilities of the generators, grid operators, end-users and electricity market stakeholders in an electricity system, with the aim of operating all parts as efficiently as possible, minimising costs and environmental impacts, and maximising system reliability, resilience and stability.
- **Total primary energy supply (TPES)** refers to indigenous production and imports of energy, adjusted for exports, international marine bunkers and international aviation bunkers and stock changes (if positive).

Agriculture and forest

- **Agroforestry** refers to the practice of mixing production systems to increase land productivity and efficiency of resources. Agroforestry can also contribute to enhance carbon sequestration.
- **Enteric fermentation** refers to the fermentation of feed as part of the normal digestive processes of livestock, resulting in methane emissions.
- **Manure management** refers to the use and treatment of manure from livestock (which can produce methane) in a sustainable manner.

Policy instruments

- **Competitive bidding procedures** include auctions and tendering schemes for renewable energy sources that allocate financial support to projects, usually on the basis of the cost of electricity production. The support granted to the winning bids can take the form of feed-in tariffs, feed-in premiums, capacity payments, certificate prices or investment grants.
- **Feed-in tariffs** offer low-carbon power producers a power purchase agreement with a fixed price per unit of electricity delivered. The fixed price is generally above the wholesale electricity price. Feed-in tariffs are typically coupled with guaranteed access to the grid.
- **Feed-in premiums** are similar to feed-in tariffs. They enable low-carbon power producers to sell electricity at wholesale prices, with a premium added to the wholesale price to compensate for higher costs. Premiums can be set as fixed premiums (a fixed amount is added to the market price for a certain period of time) or as flexible premiums (the exact amount is dependent from other criteria, e.g. market price, electricity demand, defined cap, defined floor).
- **Obligations and mandates** require a level of renewable energy in some capacity such as building standards/regulations, biofuel blending, renewable energy installations in new construction, etc. For instance, blending mandates requires Parties to meet a minimum target such as a share of total transportation fuel.
- **Regulatory standards** are applied to enforce the provisions of legislation. Standards contain performance requirements and can be applied to vehicles, power plants, energy-using equipment and buildings.
- **Renewable portfolio standards** are obligations placed by a government on a utility company, group of companies or consumers to provide or use a predetermined minimum renewable share of installed capacity, or of electricity or heat generated or sold. A penalty can apply for noncompliance. These policies are also known as "renewable energy quotas," "renewable electricity standards," "renewable obligations," and "mandated market shares," depending on the jurisdiction.
- **Tax incentives** are fiscal instruments to promote investment in energy efficiency or in renewable energy. These include investment tax credits, production tax credits and reductions in taxes on sales, energy, carbon, excise, value added, etc.

Note

1. OECD (2015), *Overcoming Barriers to International Investment in Clean Energy*, Green Finance and Investment, OECD Publishing, Paris, *http://dx.doi.org/10.1787/9789264227064-en*.

ORGANISATION FOR ECONOMIC CO-OPERATION AND DEVELOPMENT

The OECD is a unique forum where governments work together to address the economic, social and environmental challenges of globalisation. The OECD is also at the forefront of efforts to understand and to help governments respond to new developments and concerns, such as corporate governance, the information economy and the challenges of an ageing population. The Organisation provides a setting where governments can compare policy experiences, seek answers to common problems, identify good practice and work to co-ordinate domestic and international policies.

The OECD member countries are: Australia, Austria, Belgium, Canada, Chile, the Czech Republic, Denmark, Estonia, Finland, France, Germany, Greece, Hungary, Iceland, Ireland, Israel, Italy, Japan, Korea, Luxembourg, Mexico, the Netherlands, New Zealand, Norway, Poland, Portugal, the Slovak Republic, Slovenia, Spain, Sweden, Switzerland, Turkey, the United Kingdom and the United States. The European Union takes part in the work of the OECD.

OECD Publishing disseminates widely the results of the Organisation's statistics gathering and research on economic, social and environmental issues, as well as the conventions, guidelines and standards agreed by its members.

OECD PUBLISHING, 2, rue André-Pascal, 75775 PARIS CEDEX 16
(91 2015 06 1P) ISBN 978-92-64-23267-9 – 2015

www.ingramcontent.com/pod-product-compliance
Lightning Source LLC
Chambersburg PA
CBHW061222240926
51846CB00006B/1315
9789264232679